SCIENCE AND THE TRINITY

Science and the Trinity

The Christian Encounter with Reality

JOHN POLKINGHORNE

Yale University Press   New Haven and London

Published 2004 in the United States by Yale University Press and in
Great Britain by SPCK Publishing.

Scriptural quotations throughout this work are taken from the
New Revised Standard Version (NRSV).

Designed by James J. Johnson and set in Janson type by Tseng Information
Systems Inc., Durham, North Carolina.
Printed in the United States of America.

Library of Congress Cataloging-in-Publication Data
Polkinghorne, J. C. 1930–
Science and the Trinity : the Christian encounter with reality /
John Polkinghorne.
p. cm.
Includes bibliographical references and index.
ISBN 0-300-10445-6 (alk. paper)
1. Religion and science.   2. Trinity.   I. Title.
BL240.3.P65   2004
261.5′5—dc22   2004046947

A catalogue record for this book is available from the British Library.

The paper in this book meets the guidelines for permanence and durability
of the Committee on Production Guidelines for Book Longevity of the
Council on Library Resources.

ISBN 978-0-300-11530-7 (pbk. : alk. paper)
ISBN 0-300-11530-X (pbk. : alk. paper)

10 9 8 7 6 5 4 3 2 1

To
Eric Hutchison
Mary Tanner
Michael Welker

Contents

Acknowledgements

The substance of this book derives from the Warfield Lectures that I gave at Princeton Theological Seminary in March 2003, under the title 'Trinitarian Perspectives'. I am grateful to the Seminary and its President, Thomas Gillespie, for the invitation to give the lectures and for generous hospitality received while doing so. I have also benefited from a number of helpful comments made to me by staff and students at the Seminary.

I worked on a draft of the lectures during a two-month visit to the Wissenschaftlich-Theologisches Seminar in the University of Heidelberg. I am most grateful to my friend Professor Michael Welker for his hospitality and for his insightful comments on the material of this book, though the responsibility for what is actually written remains, of course, with me. I thank the Alexander von Humboldt Foundation for the award of a *Forschungspreis* and the Rektor of the University of Heidelberg for hospitality.

I also want to thank Simon Kingston and Joanna Moriarty of SPCK for helpful suggestions in relation to a draft of the manuscript, and my wife Ruth for her help in correcting the proofs.

I dedicate this book to three friends who have been particular sources of help and encouragement to me on my theological journeyings. Eric Hutchison led a Bible study group that I attended for several years. His teaching helped me to learn how one may explore the scriptures in a way that is open to their spiritual power and insight in a manner that expands one's thinking, rather than narrowing it. Mary Tanner continued my biblical education by teaching me Hebrew and Old Testament at Westcott House. We have remained good friends and I have enjoyed many discussions with her, particularly on ecumenical matters. Michael Welker and I first met at the Center of Theological Inquiry at Princeton. Over the subsequent years we have been friends and collaborators together in theological exploration. Michael has helped an Anglican scientist like me to enter into something of the riches of Reformed systematic theology. These three friends have been inspirations to me and I am glad to have this opportunity of acknowledging publicly my debt of gratitude to them.

Preface

Significant scientific advances often begin with the illuminating simplicity of a basic insight (for example, Mendeleev's arrangement of the chemical elements in the recurrent patterns of the periodic table), but they persist and persuade through the detailed and complex explanatory power of subsequent technical development (understanding the mutually interactive properties of atoms in terms of their outer electron-shells). To take another example, the generation of biologists that followed the publication of *The Origin of Species* argued about Darwin's ideas, and some attempted in various ways to modify the notion of natural selection. The dust only really began to settle in the twentieth century, when the rediscovery of Mendel's ideas concerning the inheritance of characteristics, and the subsequent fusion of genetics and evolutionary thinking, led to the synthesis of neo-Darwinism. This yielded a theory of sufficient complexity to be broadly persuasive to virtually all biologists (all accept the concept of descent with modification, though there are contemporary scientific debates about whether natural selection by itself is the totally sufficient causal story). In science, it is well-articulated proposals that lead ultimately to conviction.

What is true about the scientific understanding of the physical and biological world is also true of theology's quest for an understanding of God. Broad general ideas are attractive (a divine Mind behind the order of the universe), but I believe that theism only becomes truly persuasive when it is elaborated in greater detail and when it is anchored in the experience and interpretation that are preserved and propagated within a religious tradition. That is why there are very many members of faith communities, but very few free-standing philosophical theists.

Core theological activity, like core scientific activity, has to make the intellectual effort to work out its understanding in terms as complete and as detailed as are possible for it to achieve. This interpretative task is obviously much more difficult when its subject is the God who transcends humanity, rather than the physical world that we transcend. Yet, when it comes to the dialogue between science and religion, the discussion has frequently been conducted in rather general terms. One result of this has been that the agenda has often been set from the science side. Topics such as the discoveries of modern astronomy, quantum theory, evolutionary biology, genetics and neuroscience have regularly provided the heads of discussion. Of course there is value and validity in this approach, but it by no means represents the only way in which the issues can be considered.

Like my fellow scientist-theologians, I have often operated in this science-led mode. For example, I have written and spoken about natural theology, using the deep intelligibility of the universe and the finely tuned fruitfulness of its history as guides to the organisation of the argument. As a quantum physicist, I have been concerned with how that subject's veiled account of created reality might influence theology's understanding of God's relationship to the physical universe. Yet I have also wanted to make clear, as opportunity offered, that the central source of my own belief in God does not lie in such matters. Rather, it is to be found in my encounter with the figure of Jesus Christ, as I meet him in scripture, in the Church and in the sacraments. For me, it is Trinitarian belief that is truly persuasive belief. Of course, that belief is much more complex than simple recognition of the Mind of God behind the order of nature, just as modern quantum theory is more complex than Max Planck's original idea that energy comes in packets. Yet, Trinitarian belief is complex in ways that seem to me to be necessary to match the depth of experience and insight recorded in the Bible, and continued in the ongoing life of the Church.

The aim of this volume is to make a contribution to the science and religion dialogue in which it is largely theological concerns that are allowed to shape the argument and to set the agenda. After an introductory survey of different approaches to the dialogue between science and theology (Chapter 1), the heads of discussion are scripture (Chapter 2), a theology of nature (Chapter 3), the nature of God (Chapter 4), sacrament (Chapter 5), and eschatological hope (Chapter 6).

I believe that a discussion of this kind has to be undertaken from the standpoint of a particular faith tradition (see also the Postscript, Chapter 7). It is only as the world faiths articulate their own positions with as much clarity as possible that the subsequent step of honest and difficult dialogue between them can really begin. Since I am a Christian, the perspectives that I adopt are Trinitarian. Because I am a theoretical physicist, the style of thinking I adopt is a 'bottom-up' approach, which seeks to move from experience to understanding. The purpose of the discussion is to explore and defend the thesis that Christian belief is as illuminating and intellectually credible in the twenty-first century as it was in the century that gave it birth, and as it has been, in my opinion, over the intervening centuries.

In the first chapter the forms of active dialogue between science and theology are reconsidered. Previous attempts at classifying these approaches have been framed in terms of an analysis of the styles of mutual encounter between the two disciplines. Lately, as the topics of conversation have become increasingly theological in their scope and significance, it has seemed to me necessary to move to schemes based on theological content. Four approaches are therefore categorised: Deistic, Theistic, Revisionary and Developmental, with Paul Davies, Ian Barbour, Arthur Peacocke and the present writer chosen as prototypical examples. The hope is expressed that more sustained participation in the dialogue by mainline theologians will lead to the development of a fifth approach, one that might come to be called Systematic.

The second chapter turns to the use of scripture. Scientists sometimes feel that Christian theologians appeal to the Bible in order to foreclose discussion rather than to facilitate it. It seems important to establish at the start that scripture does not function in this mind-closing way. On the contrary, for though the Bible is indeed an indispensable and authoritative source for the Christian, it is one that must be approached in ways that are subtle and complex rather than literal and unproblematic. Scriptural roles include the evidential, the spiritual and the contextual. In a respectful reading of the Bible attention must be paid to genre and to historical authorial setting, together with acceptance of progressive spiritual and theological development over the many centuries of the writings' compilation. Honesty requires the acknowledgement of the presence of unedifying passages in the Bible, and hermeneutic adequacy requires recognition of the polysemous character of the texts, capable of conveying meaning at several different levels. A brief consideration of Paul's use of the Hebrew scriptures shows that it is characterised both by respect and by a creative freedom of interpretation and use.

Chapter 3 essays a theology of nature based on a Trinitarian meta-interpretation of science's account of cosmic process. Of particular significance are the widespread relationality characteristic of modern physics, the veiled nature of quantum reality, and the open and information-generating character of the behaviour of complex systems. These features are seen to be fully consistent with the belief that the Trinitarian God is the Creator of the universe.

In the fourth chapter, it is argued that because of its richness of insight, a 'thick' Trinitarian theology is more persuasive than is the case for a less elaborated form of theistic belief. Panentheism is rejected because of its unhelpful blurring of the distinction between Creator and creation. All that is needed by way of correction to classical theology is the recovery of a strong account of divine immanence. An approach to Trinitarian understanding 'from below' is commended on the basis of a theological realism that starts with the record of the Christian community's experiential encounter with what theologians call the economic Trinity, that is to say God known through revelatory acts. Divine unity is held not to require an excessive doctrine of divine simplicity. In particular, it is proposed that there is an eternal-temporal polarity within the divine nature and that God's perfection lies in an ever-appropriate relationship to changing creation, rather than in an absolute divine immutability. It is tentatively suggested that it would be sufficient for the fulfilment of strong Christological and soteriological criteria if it is the temporal pole of the Second Person that is involved in the historical episode of the incarnation.

Chapter 5 considers the experience of the faithful Christian believer that centres on the Eucharistic worship of the Church. This is seen as being a fulfilment of the Lord's command to do this in remembrance of him and emphasis is placed on the total action of the visibly gathered Christian community meeting in the presence of the invisible risen Christ. Thanksgiving for creation, remembrance of the cross and anticipation of the final vindication of the Lordship of Christ, together with the invocation of the Spirit, combine in the Trinitarian worship of the Church. This chapter is the one that makes the least direct appeal to science itself, but its approach is a theological counterpart to the scientific strategy of grounding understanding in interpreted experience.

The sixth chapter takes up the eschatological themes of a destiny beyond death, both for human individuals and for the whole created universe. A hope of this kind is held to have no natural basis, for it can be founded only on the faithfulness of God. The credibility of this hope is explored in terms of four eschatological criteria, with particular emphasis on the necessary blending of both continuity and discontinuity in the transition from the old creation to the life of the new creation, a final reality already partially in existence and stemming from the seminal event of Christ's resurrection. It is proposed that the human soul may best be understood as the immensely complex information-bearing pattern carried by the body, a modern version of the Aristotelian-Thomistic concept of the soul as the form of the body. Eschatological hope both depends upon Trinitarian belief and also completes the credibility of that belief.

The final chapter (whose title I have borrowed from Søren Kierkegaard), offers a defence of the particularist stance that is taken in this book.

CHAPTER ONE

Four Approaches to the Dialogue between Science and Theology

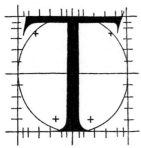HEOLOGY seeks to speak of God, the One who is the source of all created being. Therefore, to some degree theology must take account of all forms of truth-seeking investigation into the nature of what is. Among such enquiries, the discoveries of science are of clear significance as they tell us about the pattern and history of the universe. Theology, in its turn, regards that universe as being God's creation. It is therefore not surprising that the dialogue between theological thinking and the science of the day has had a long history. One may recall that Augustine, in his *Literal Commentary on Genesis*, does not at all produce the kind of flat-footed 'creationist' seven-day account that his title might suggest to a modern reader but, among other things, emphasises that if well-supported contemporary understanding of the natural world appeared to conflict with a traditional interpretation of scripture, then that interpreta-

tion should be reconsidered in the light of this other knowledge.

The form that this interaction between science and theology has taken has varied considerably over the centuries. Certainly it is not possible to subsume it under some single simplistic rubric, such as the 'warfare' of the scientific with the theological or their harmonious conflation with each other. John Hedley Brooke, in his scholarly survey covering the period from the seventeenth century to the end of the nineteenth century,[1] has made clear the variety and the complexity of the historical interaction between science and religion. For the general enquirer, careful consideration of these matters has been hampered by the assiduous propagation of the myth of a battle between scientific light and religious darkness—a misrepresentation particularly popular in those sections of the media that love stories of confrontation.

Someone who has suffered from a campaign of misrepresentation in this respect has been John Calvin. It has become a cliché to quote, as an egregious example of theological blindness to scientific truth, his alleged remark 'Who will place the authority of Copernicus above that of the Holy Spirit?' In actual fact this remark is nowhere to be found in Calvin's writings. It appears to have originated as a plain invention by the nineteenth-century writer F. W. Farrar, who (I regret to have to say) was the Dean of Canterbury Cathedral.[2] The injustice is particularly great since Calvin's idea of accommodation—the concept that the writers of scripture expressed

1. J. H. Brooke, *Science and Religion*, Cambridge University Press, 1991. P. J. Bowler, *Reconciling Science and Religion*, University of Chicago Press, 2001, carries the story into the early twentieth century with similar conclusions.

2. D. Alexander, *Rebuilding the Matrix*, Lion Publishing, 2001, p. 129.

themselves in ways that would be accessible to the common reader of their time—led him to warn expressly against treating the Bible as a quarry from which to attempt to hew scientific conclusions. Calvin wrote in his *Commentary on the Psalms* that 'The Holy Spirit has no intention to teach astronomy . . . the Holy Spirit would rather speak childishly than unintelligibly to the humble and unlearned'.[3] To continue this Anglican acknowledgement of Reformed good sense in thinking about these matters, let me also note that Benjamin Warfield (who founded the lectures that I had the honour to give), writing in the aftermath of what is mythically represented as a paradigm period of great conflict between science and religion, said that, in his opinion, he 'did not think that there is any general statement in the Bible or in any account of creation, either as given in Genesis 1 and 2 or elsewhere alluded to, that need be opposed to evolution'.[4] Finally, we should gratefully recall that one of the leading contemporary theologians to take a sustained interest in how science and theology relate to each other is also from the Reformed tradition. I refer, of course, to Thomas Torrance, whose writing often refers to scientific matters, particularly those associated with his two great scientific heroes, James Clerk Maxwell and Albert Einstein.[5]

Recent years have seen a very active engagement in the dialogue between science and theology, mostly conducted by those whose intellectual formation has been on the scientific side of the border. It is quite customary to date this vigorous

3. Quoted ibid., p. 132.
4. Quoted ibid., p. 201.
5. T. F. Torrance, *Theological Science*, Oxford University Press, 1969; *Divine and Contingent Order*, Oxford University Press, 1981; see also J. C. Polkinghorne, *Faith, Science and Understanding*, SPCK/Yale University Press, 2000, § 8.3.

activity from the publication in 1966 of Ian Barbour's *Issues in Science and Religion*,[6] and there is no doubt that this was indeed a seminal event in terms of its wide influence in the academic world. However, many of the isssues raised in Barbour's book had been anticipated in Eric Mascall's Bampton Lectures of 1956, *Christian Theology and Natural Science*,[7] where their treatment was heavily influenced by the author's Thomistic style of thinking.

Since those days, the mutual conversation between science and theology has intensified and the rate of relevant publication has quickened considerably. The resulting dialogue has proved to be a kind of spiral process, circling ever inwards to deeper engagement with topics of central concern to Christian theology.[8] There are certain natural frontier issues, such as the doctrine of creation, the status of natural theology, and the critique of a crassly reductive physicalism, that will always engage the attention of workers in this field. Yet much of what can be said in respect to these issues is as consistent with the distant God of deism as it is with the God of providential theism who truly interacts with the history of creation. The recognition that this is so has had the effect in the last ten years of bringing to the top of the agenda a more demanding question. It asks in what way one might hope to understand divine providential action to be exercised in the kind of world whose processes are described by the orderly accounts that science seems to offer. No fully agreed consensus has emerged

6. I. G. Barbour, *Issues in Science and Religion*, SCM Press, 1966.

7. E. L. Mascall, *Christian Theology and Natural Science*, Longman, 1956.

8. For discussion of some of the contributors to this dialogue, see J. C. Polkinghorne, *Scientists as Theologians*, SPCK, 1996; *Faith, Science and Understanding*, ch. 8.

from these discussions, but it has been widely recognised that the intrinsic unpredictabilities that twentieth-century physics has uncovered as limits on our knowledge of detailed behaviour, both in quantum theory and in chaos theory, have significantly qualified the kind of merely mechanical account of physical process that previously had seemed to be the deliverance of science.[9] As a result, an honest appeal to science cannot be used to discredit belief in God's providence acting within the divinely ordained open grain of nature. Moreover, if creatures can act as agents in the world (a capacity that human beings directly experience but which itself is not, as yet, well understood in terms of a scientific account of detailed process), it would not seem reasonable to deny the possibility of some analogous capacity in the Creator.

Recently there has been a further twist in this spiralling engagement of science and theology, in that issues of eschatological credibility have become matters of current discussion. Here the main initiative must lie with theology but science can pose, with some sharpness, some of the questions that need to be addressed and it can even, to a minor degree, constrain the form of some of the answers that can be proposed.[10]

One consequence of this increasingly more specific engagement with topics of central theological concern has been to show up more clearly that there are some significant differ-

9. See R. J. Russell, N. Murphy and A. R. Peacocke (eds.), *Chaos and Complexity*, Vatican Observatory, 1995; J. C. Polkinghorne, *Belief in God in an Age of Science*, Yale University Press, 1998, ch. 3. The word 'intrinsic' is important here. No attempt is being made to revive the discredited concept of 'the God of the gaps', supposed to operate within the lacunae of current scientific knowledge and so always liable to vanish with the further advance of that knowledge.

10. J. C. Polkinghorne and M. Welker (eds.), *The End of the World and the Ends of God*, Trinity Press International, 2000; J. C. Polkinghorne, *The God of Hope and the End of the World*, SPCK/Yale University Press, 2002. See also Chapter 6.

ences of approach to the dialogue between science and theology that are present in the thinking of various participants in the conversation. When the principal matters under consideration were the deep intelligibility of the physical world, the anthropically fruitful history of the universe, the evolutionary exploration of inherent potentiality, and the inadequacy of a reductionist theory of quarks and gluons to fulfil the grandiose claim to be a Theory of Everything, it was comparatively easy to discern a considerable degree of unanimity among those who sought to incorporate these insights into a theology of nature. When the matters under consideration came to include such topics as divine providential engagement with the specificities of history, the significance of human personhood, the status of Jesus Christ, and the hope of a destiny beyond death, then much more diverse assessments began to be made concerning how relevant and constraining are scientific conclusions, and how appropriate is a scientific style of thinking, for the theological task of the discussion of these issues. In the light of these developments, I want to re-examine the range of approaches that have come to be pursued in the contemporary dialogue between science and theology.

In his Gifford Lectures,[11] Ian Barbour offered a taxonomy of the different ways in which he saw that it had proved possible to relate science and religion. His scheme has become something of a classic grid which has been used by many subsequent writers on the subject. Barbour's fourfold classification employs the headings of Conflict, Independence, Dialogue and Integration.[12] Conflict corresponds to the uncom-

---

11. I. G. Barbour, *Religion in an Age of Science*, SCM Press, 1990, ch. 1.

12. This is worked out in relation to specific topics in I. G. Barbour, *When Science Meets Religion*, HarperCollins, 2000.

promising choice demanded by those who believe that either science or religion must be the sole occupant of the intellectual driving seat. The contradictory stances of scientism and creationism (the latter word understood in its curious North American literalist sense) meet here in agreeing that a choice has to be made that will then commit one to being wholly intolerant of any other point of view. Independence presents us with a far less drastic option. Science and religion are considered to use different languages, to pose different questions, to consider different dimensions of experience, and generally to operate insulated from one another. This division is quite often presented in terms of a dichotomy that separates the domains of public knowledge (science) and private opinion (religion), or by way of a similar distinction between a concern with facts and a concern with values. Independence is a popular stance for scientists who do not want to dismiss religion altogether, but who equally do not want to worry very much about its truth claims.[13]

The stances of Dialogue and Integration both take a much more positive view of the possibility of fruitful exchange between science and theology. The former believes that the two disciplines have things to say to each other. For example, both offer insights into the nature of cosmic history. Their perspectives are different and there is no direct entailment between them, but nevertheless one can reasonably expect the two sets of insights to exhibit some degree of compatibility with each other. It is in this mode that many would consider that the idea of evolutionary process and the concept of continuous creation can be seen as mutually enlightening. Inte-

13. See, for example, S. J. Gould, *Rock of Ages*, Ballantine, 1999.

gration seeks a much closer degree of engagement, such as would be proposed, for instance, in the synthetic thinking of Teilhard de Chardin,[14] or in the metaphysical scheme of process thought.[15]

Most serious contributors to the dialogue between science and theology reject both the head-on collision of Conflict and the mere talking past each other of Independence. Both of these approaches are seen as being either inadequate or misleading. Attention, therefore, has concentrated on the mediating ground of interactive encounter. In his own thinking, Barbour acknowledges that he uses a combination of the two stances of Dialogue and Integration that he has described.[16] Other writers have sought to delineate the frontier exchange in somewhat different ways.

John Haught produced an alliterative scheme using the concepts of Conflict, Contrast (similar to Independence), Contact and Confirmation.[17] The stance of Contact acknowledges that science and theology interact with each other and that, in consequence, new scientific discoveries can exert an influence on theological thinking. An instance would be the impact that biological evolution and Big Bang cosmology have had on the way that theologians talk about the doctrine of creation. The stance of Confirmation makes the bolder claim that 'religion, without in any way interfering with science, paves the way for some of its ideas, and even gives a special kind of blessing . . . to the scientific quest for truth'.[18] This possibility

14. P. Teilhard de Chardin, *The Phenomenon of Man*, Collins, 1959.

15. A. N. Whitehead, *Process and Reality*, Free Press, 1978; P. A. Schilpp (ed.), *The Philosophy of Alfred North Whitehead*, Open Court, 1971.

16. Barbour, *Religion in an Age of Science*, p. 30.

17. J. Haught, *Science and Religion*, Paulist Press, 1995.

18. Ibid., p. 4.

might be illustrated by the claim, made by some historians of ideas,[19] that it was the Judaeo-Christian-Islamic concept of a creation whose order had been freely chosen by its rational Creator that provided an important element in the intellectual setting that enabled modern science to come to birth in Europe in the seventeenth century.

In my turn, I have sought to redescribe Barbour's two forms of constructive encounter in terms of the categories of Consonance and Assimilation. The former refers to the way in which 'science does not determine theological thought but it certainly constrains it' by conditions of mutual congruence. In contrast, the category of Assimilation refers to attempts 'to achieve a greater merging of the two disciplines'.[20] I am suspicious of this latter approach, since I believe that it tends to result in science playing too great a controlling role in the proposed convergence, with the result that there is a danger that theological concerns become subordinated to the scientific. I fear this effect much more than Barbour does, and hence my choice of a less complimentary term to describe the synthetic exercise.

More recently, a group of younger scholars came together with the intention of formulating a revisionary approach to these issues. They were influenced by what they saw as the postmodern state of cognitive pluralism, with its suspicion of all attempts at an overarching meta-narrative. For these reasons, the group sought to go beyond the ideas of the scientist-theologians of my generation. Appropriately enough, no single agreed theme emerged and their joint volume expounds

---

19. See, for example, S. Jaki, *The Road of Science and the Ways to God*, Scottish Academic Press, 1978; C. A. Russell, *Cross-Currents*, Inter-Varsity Press, 1986.

20. Polkinghorne, *Scientists as Theologians*, pp. 6-7.

a variety of contrasting views.[21] The spread of the options offered is wide, ranging from Willem Drees's reliance on scientific naturalism, which only permits theology a peripheral role as the possible source of answers to limit questions, to the contributions of the two editors, Niels Gregersen and Wenzel van Huyssteen, who present less drastic proposals, based on the concepts of contextual coherence and post-foundational rationality respectively. The former approach uses coherence, evaluated within a general web of beliefs and knowledge and exercised in a pragmatically effective way, as its critical norm, while the latter envisages a flexible, multi-dimensional concept of rationality, located within the context of living and developing traditions. The total offering of the six contributors is of considerable interest, but it is too diverse for short summary. I have to say that personally I remain persuaded of the validity of a carefully nuanced critical realism in both science and theology. It seems to me that a number of the points made by its critics are more in the nature of an exploration of what might be involved in understanding the qualifier 'critical', rather than amounting to a negation of the concept itself. As the present volume illustrates, I am unwilling to relinquish the grand scheme of Trinitarian theology, anchored in the narratives of the canonical tradition.

I have come to believe that the increasing theological sophistication of the interaction between the two disciplines means that a different kind of classification is now needed, making somewhat finer distinctions that relate not only to the methods of discourse employed but also to the content of what

21. N. H. Gregersen and W. J. van Huyssteen (eds.), *Rethinking Theology and Science*, Eerdmans, 1998.

is allowed to enter into the mutual conversation. The principal purpose of this chapter is to explore a fourfold discrimination of the different kinds of positive interaction that have actually been taking place. In the course of the discussion I shall make illustrative use of the ideas of four contributors to the science-theology dialogue whose work seems, respectively, to fall into these four categories. The scheme I propose is essentially theological, rather than methodological, in its character. The first of its categories is:

(1) *Deistic.* The adjective is justified because the degree of engagement between science and theology that is envisaged in this approach is modest, with the initiative coming mainly from the scientific side. The core of the strategy is the recognition that there are significant questions that arise from the experience of doing science, but their character is such that answering them requires a move outside of science itself. If these meta-questions are to be addressed, the resource for doing so must therefore be more than purely scientific. It is then suggested that the concept of some form of Cosmic Intelligence, of the kind that the God of deism would represent, is a rationally coherent possibility that should be taken into account by those who are seeking a maximal degree of understanding.

The questions that give rise to this kind of argument generally centre on two topics. One is the character of the laws of nature. From a purely scientific standpoint, these are the scientific givens that are discovered to be acting in the universe and which serve to define the character of its physical fabric. Science does not explain their origin but simply uses them as the basis for its discussion of the detailed character of physical and biological process.

Every discipline has to rest on an unexplained foundation. For science this is provided by the fundamental laws of nature, just as theology rests on the given existence of the deity conceived, for example, as a maximal necessary being. Nothing comes of nothing, and no explanatory scheme can be totally self-explanatory, as if it were totally free from any unexplained input. In forming our view of reality, David Hume would encourage us to take our stand on the ground that treats the given properties of matter as the basis for all explanation. However possible such a materialist stance might have seemed in the eighteenth century, at the beginning of the third millennium, physical scientists, in particular, are apt to discern in the laws of nature a character that is highly suggestive that there is more to learn about them than unaided science can find the means to say. Hume's scientistic strategy appears unappealing. A move beyond science then becomes a possibility to be explored (see also Chapter 3).

For one thing, there is the fact of the profound intelligibility of the universe that has made its laws actually accessible to us, so that whether it is quantum theory's account of the subatomic world, or the account that general relativity gives of the vast domains of cosmic curved spacetime, we are able to understand regimes that are remote from everyday experience and whose character demands highly counterintuitive ways of thinking if we are properly to comprehend them. Moreover, mathematics—that apparently most abstract of human activities—turns out to provide us with the key to unlock these physical secrets. And it is not just any old mathematics that fulfils this revelatory task, but the kind whose equations are endowed with the unmistakeable character of mathematical beauty. The fundamental structure of the universe is astonishingly rationally transparent to us, thereby affording science the possibility of making its discoveries. The universe is also rationally beautiful, thereby affording scientists the reward of wonder as the recompense for all their demanding labour. Are all these matters just our luck, or are they signs that there is a divine Mind that lies behind the marvellous order of the cosmos?

Meta-questioning of the laws of nature does not end there, for there is also the remarkable fact that it is their quantitative specificity that alone has allowed the development of carbon-based life in the course of cosmic history. While that history has been characterised by evolutionary exploration of potentiality, the process only resulted in such fruitful consequences as human beings because it took place in a context that was, so to speak, 'finely tuned' to permit the possibility of carbon-based life developing. The detailed character of the physical fabric of the world was of critical importance and had to be tightly constrained. There has been much discussion of what should be made of these Anthropic Principle considerations, but they certainly raise metaphysical questions to which belief in God can provide, at the least, an economic and coherent answer.[22]

The second topic that has attracted much attention has been the coming to be of self-conscious life, surely one of the most remarkable developments known to us in the fourteen-billion-year history of the cosmos. In the dawning of rational self-awareness, the universe began to come to know itself, and thereby science became a future possibility. To many people this does not look like just an incredibly happy accident.

22. See J. D. Barrow and F. J. Tipler, *The Anthropic Cosmological Principle*, Oxford University Press, 1986; J. Leslie, *Universes*, Routledge, 1989.

One contributor to the dialogue between science and theology whose thinking has been almost entirely confined to these issues is Paul Davies.[23] His thought is largely science-driven and theological considerations play a distinctly subordinate role in it, a fact that enables Davies to make his somewhat notorious comment that, in his opinion, 'science offers a surer path to God than religion'.[24] It is, of course, the etiolated God of deism—the Cosmic Architect or the Great Mathematician—who is, at best, the endpoint of this kind of argument. *God and the New Physics* concludes with the picture of a deity who is a kind of ingenious demiurge.[25] In a later and more developed book, *The Mind of God*, Davies admitted that it was not obvious to him that 'this postulated being who underpins the world has much relationship to the personal God of religion, still less the God of the bible or the Koran'.[26] One is reminded of Albert Einstein, who liked to use talk about God as a kind of cipher for the rational order of the universe, but who vigorously repudiated belief in a personal God, saying that if he had a God it was the (pantheistic) God of Spinoza.[27]

Davies ends *The Mind of God* by stating,

> I cannot believe that our existence in this universe is a mere quirk of fate, an accident of history, an incidental blip in the great cosmic drama. Our involvement is too intimate. . . . Through conscious beings the universe has generated self-awareness. . . . We are truly meant to be here.[28]

23. P. Davies, *God and the New Physics*, Dent, 1983; *The Mind of God*, Simon and Schuster, 1992.

24. Davies, *God and the New Physics*, p. ix.

25. Ibid., ch. 17.

26. Davies, *The Mind of God*, p. 191.

27. M. Jammer, *Einstein and Religion*, Princeton University Press, 1999.

28. Davies, *The Mind of God*, p. 232.

Davies' thinking illustrates oth the possibility of a stand-alone natural theology and also its theological thinness. Thomas Torrance is surely right to insist that natural theology must be integrated with the rest of the discipline of theology in the single search for the knowledge of God, if the theological enterprise is to prove to have sufficient richness and depth.[29] All possible resources for insight must be drawn into play. It is not enough to recognise the significance of persons simply in terms of their self-awareness; they must also be regarded as perceivers of value and as participants in the religious encounter with the reality of the sacred. There is also a dark side to human nature that must be acknowledged, that corruptive influence that theologians consider under the category of sin.

The God of deism ultimately proves too diminished a deity for the question of that kind of divine existence really to seem to matter all that much. Hence the decay that one can trace historically in the course of eighteenth-century rational religion, moving from natural theology to natural philosophy and then on to atheism itself.[30]

(2) *Theistic.* In actual fact, almost all believers in God are adherents of a particular faith tradition, though no doubt with a variety of degrees of commitment to the core of their tradition. The individual person of general but unanchored theistic inclinations, such as Paul Davies, is a rather unusual case. Most belief stems not only from metaphysical argument but also from the experience of worship and practice within a religious community. Of course, this observation immediately raises all the perplexities inevitably associated with the great diversities

29. T. F. Torrance, *Theological Science* and *Reality and Scientific Theology*, Scottish Academic Press, 1985.
30. M. Buckley, *At the Origins of Modern Atheism*, Yale University Press, 1987.

of belief exhibited by the world faith traditions in their accounts of the nature of the sacred. Here is both a problem and a potential for future theological discourse—a problem because of the apparent cognitive dissonance, and a potential because of the rich diversity of insight and experience expressed and preserved. I regard this interfaith issue as one of the most important subjects on the theological agenda at the start of the third millennium,[31] and I shall have a little more to say about it in the Postscript. Meanwhile I must confine my discussion to what has been going on within Christianity.

The ways in which a religious tradition impinges upon theological reflection can be quite diverse. While the believing community will provide the general context for theological discourse, there are a variety of manners in which specific issues might be engaged. This section, and the two sections that follow, illustrate this sort of diversity within the context of a Christian concern for the issues of science and religion.

By a theistic approach pursued within the Christian context, I mean one that is genuinely influenced by a Christian style of thought and experience, but which does not necessarily come to grips with a fully articulated range of doctrinal issues, such as those that are laid out in the articles of the Nicene Creed. A theistic approach of this kind will not at all be content with natural theology alone. It will draw inspiration from the Bible, and in particular from the life and words of Jesus of Nazareth. It will be concerned with how one may understand divine providential action to be exercised in the

31. See J. C. Polkinghorne, *Science and Christian Belief/The Faith of a Physicist*, SPCK/Princeton University Press, 1994, ch. 10; *Scientists as Theologians*, ch. 5; *Science and Theology*, SPCK/Fortress, 1998, ch. 7.

history of the world and how one may understand the practice of petitionary prayer, but it may, in the end, be somewhat reserved about exactly what responses should be given to these questions. It will be a theological activity that is influenced by the experience of worship and which values the collective insights of the Christian community, but it will allow itself a good deal of freedom about how it actually makes use of the resources of the tradition.

My example of someone who operates largely in a theistic mode is Ian Barbour. No one could fail to see that his thinking is contained within the envelope of Christian understanding. He makes use of a variety of general biblical concepts and in particular he lays stress on the cross of Christ as the outstanding exemplification of the power of suffering love. For Barbour, relationship and history are the most important categories for theological discourse and he is much influenced by the inspirational Christology of Geoffrey Lampe.[32] In Barbour's view, 'What was unique about Christ, in other words, was his relationship to God, not his metaphysical "substance" '. He suggests 'that in an *evolutionary perspective* we may view both the human and divine activity in Christ as a continuation and intensification of what had been occurring previously. We can think of him as representing a new stage in evolution and a new stage in God's activity'.[33] This implies, as Barbour acknowledges,[34] that what is special about Jesus is a matter of degree. One might say that Jesus was better at being truly human than the rest of us have succeeded in being. If that is a fair way of putting it, it would seem that Jesus is

32. G. W. H. Lampe, *God as Spirit*, Oxford University Press, 1977.
33. Barbour, *Religion in an Age of Science*, pp. 210-11, his italics.
34. Ibid., p. 213.

an inspiring example but he does not act as our redeeming Saviour. Although Barbour quotes the words of Paul, 'God was in Christ reconciling the world to himself' (2 Corinthians 5:18), it seems that they are to be understood in a purely exemplificatory sense. Yet an adequate account of the soteriological role of Christ, so strongly testified to by the Church from the very first, is surely an indispensable criterion of a fully adequate Christology. Affording an example does not seem to be enough, for we stand in need of a source of the grace that will enable human beings actually to follow that example. The *imitatio Christi* is not something that we can do just on our own.

In relation to central Christian dogmas in general, Barbour is often content to summarise what has been said by others and to be somewhat reticent about what he himself thinks. This tendency is particularly notable in respect to Christ's resurrection, to which there are very few references of any kind in Barbour's corpus. For much Christian thinking, both contemporary and traditional, the resurrection is the hinge on which Christian understanding pivots. If Jesus was indeed raised from the dead that first Easter day, never to die again but to live an exalted life at the right hand of God the Father, then there is indeed something uniquely significant about him, going beyond anything that could be considered as 'a continuation and intensification of what had been occurring previously'.

The other aspect of Barbour's thinking that is particularly striking is the role that is played in it by a metaphysical understanding derived from process philosophy. He is the leading proponent in the science and theology community of that particular way of thinking. Personally, I have two principal problems with process thought. One is that its event-dominated metaphysics does not seem consonant with the physical basis provided by modern quantum theory, which assigns at least as important a role in physical process to continuous development as it does to discontinuous change, the latter occurring only intermittently at moments of measurement. My other problem is that the God of process theology is too metaphysically limited, constrained to act through persuasion alone. Such a conception of deity falls short of describing a being who could be the basis of an everlasting hope. Certainly, the process God does not seem to be the One who could have raised Jesus from the dead. Barbour is frank enough to confess that 'Process theology does call in question the traditional expectation of *an absolute victory over evil*'.[35] I also have to say that the persuasive lure of the God of process theology seems to be soteriologically insufficient, for the human condition is such that we need empowerment as well as encouragement. God must will the means as well as the desirability of our salvation. Hence the Christian concern with grace given to us (cf. Romans 7:15-25).

There is nothing at all improper in theology operating in tandem with a chosen philosophical system. One thinks of the influence of neo-Platonism on Augustine, or of the newly recovered philosophy of Aristotle on Aquinas. Yet, despite all its bold creativity, process thought does not seem to sit easily with much Christian insight into the Creator's ways with creation. That said, one must gladly acknowledge that there is a much greater richness in the theistic approach of the kind

35. Ibid., p. 264, his italics.

that I see Barbour as exemplifying than is to be found in the theologically thin account of Davies' deistic approach. There is, however, still the question of whether there might not be even richer approaches, capable of leading to deeper and more exciting conclusions, that also need to be taken into account. Two such stances remain to be considered.

(3) *Revisionary.* This third attitude corresponds to an approach which engages extensively with the range of topics that have been the traditional concern of Christian theology, but which also believes that the way in which these topics are to be treated requires radical revision in the light of modern knowledge and, in particular, as a result of scientific discovery. Such an approach will lead to much talk about Christology and about Christ's resurrection, but it will not hesitate to speak on these issues in a manner that is distinctly different from that which has been employed in the past. My representative figure for this way of treating the dialogue between science and theology is Arthur Peacocke.

Peacocke has always maintained the right of theology to speak with its own distinctive voice. He believes that theology 'refers to the most integrated level or dimension we know in the hierarchy of relations [within reality]. So it would not be surprising if the concepts and theories which are developed to explicate the nature of this activity are uniquely specific to and characteristic of it'.[36] As a very positive example of the way in which scientific understandings have influenced theological discourse, one might take the way in which Peacocke has made such helpful use of evolutionary insight to encourage a theological concept of *creatio continua* as a way of reflecting on the

36. A. R. Peacocke, *God and the New Biology*, Dent, 1986, p. 30.

process by which creatures have been allowed to explore and realise the potentiality with which creation has been endowed by its Creator.[37]

When one comes to Christology, however, a more mixed and confusing form of discourse seems to be in evidence. Peacocke sometimes speaks in terms strongly reminiscent of Barbour's way of thinking. He believes that we are encouraged 'to understand the "incarnation" which occurred in Jesus as exemplifying that emergence-from-continuity which characterises the whole process by which God is creating continuously through discontinuity'.[38] Peacocke lays great stress in his work on the transformative power of divine communication. This leads him to believe that 'what we have affirmed about Jesus is not, in principle, impossible for all humanity'.[39] Once again one faces the soteriological problem of whether the example of this 'new emergent'[40] that Jesus is felt to represent is sufficient to help those who have so far failed to emerge. It would seem that communication (*gnosis*, one might even say) is not the same as grace (participation in the divine life and energies).

Yet, at other times, Peacocke, speaking of the cross of Christ, can say that we have to conclude that '*God* also suffered with him in his passion and death'.[41] If this refers to a true divine participation in the travail of creation, it seems to

37. A. R. Peacocke, *Creation and the World of Science*, Oxford University Press, 1979, chs. 2 and 3; see also the essays in J. C. Polkinghorne (ed.), *The Work of Love*, SPCK/Eerdmans, 2001.

38. A. R. Peacocke, *Theology for a Scientific Age*, enlarged edition, SCM Press, 1993, p. 301.

39. Ibid., p. 302.

40. Ibid., p. 303.

41. Ibid., p. 310, his italics.

demand something much more like a traditional ontological understanding of the incarnation, so that in Christ God is indeed caught up in the sufferings of creation from the inside, and not simply alongside. The statement then seems to stand in unresolved tension—even contradiction, one might feel— with the language of an inspirational or functional Christology that has preceded it.

In his discussion of the resurrection, Peacocke shows a preference for the kind of language of exaltation that one finds in the Epistle to the Hebrews. He sees the resurrection appearances as 'referring to a new kind of reality hitherto unknown because not hitherto experienced'.[42] In relation to the stories of the empty tomb, however, he expresses considerable reservations. One reason for this caution is that Peacocke believes that, since in our case the molecules of our bodies will soon be dispersed and recycled after our deaths, the direct transmutation of Christ's corpse into his glorified body on the third day would serve as a poor precedent for the destiny of the rest of humanity. This supposed difficulty seems to rest on a failure to distinguish between a kind of crude resuscitatory realism that would interpret resurrection as the literal restoration of what had previously been (an idea that I too would certainly reject, not least because the atoms of our bodies have no abiding significance, for they are changing all the time), and the significance of Christ's resurrection as the wholly novel and seminal event from which God's new creation has begun to grow and which prefigures the ultimate redemption of all creation, matter as much as humanity. In my own thinking about the resurrection I attach credi-

42. Ibid., p. 281.

bility and theological importance to belief that the tomb was empty.[43]

In an earlier survey of the thinking of three scientist-theologians, I placed Peacocke between myself and Barbour in terms of the consonance-assimilation spectrum that I was then using. My judgement was that he appeared at times 'to operate in an assimilationist mode and at other times in a consonantist mode'.[44] Since then Peacocke, in an article in *Zygon* that has a strongly programmatic tone to it, has declared himself in terms that are much more unambiguously those of a radical revisionary. Beginning with the metaphor of a bridge between science and theology, he asserts that in medieval times 'one had to change vehicles halfway across the bridge as reason was left behind and the deliverances of a revealed faith took over in going from science to religion'.[45] That alleged abandonment of the usage of rational discourse seems a curious verdict on an age that was deeply concerned with questions of logic. It is also a curious verdict on the thought of someone like Thomas Aquinas, who believed that the mind seeks truth without reserve and whose dialectic method in the *Summa Theologiae* is to consider arguments both for and against the proposition being considered. At any time in Christian history, what in fact changed in crossing the bridge was not the reasonable appeal to motivated belief, but rather the kind of experience that can appropriately be invoked in seeking that motivation. The distinction made in Aquinas between reason

43. Polkinghorne, *Science and Christian Belief/The Faith of a Physicist*, pp. 115–18, 164.

44. Polkinghorne, *Scientists as Theologians*, p. 84.

45. A. R. Peacocke, 'Science and the Future of Theology: Critical Issues', *Zygon* 35 (2000), p. 120.

and faith does not relate to the acceptance or abandonment of rationality, but to the evidential sources on which a reasonable discussion can draw. It relates to the distinction between natural theology (with its reliance on general human experience) and revealed theology (with its reliance on specific acts of divine self-disclosure). In my view, the theological community has been quite as much a truth-seeking community as the community of science.

In his *Zygon* article, Peacocke then turns to the effects of the contemporary postmodernist critique. He believes that science can survive this in a way that is denied to theology. Part of his defence of this thesis lies in an appeal to evolutionary epistemology. Despite some other notable defenders of this strategy,[46] it seems to me too weak a support even for science, by itself failing to provide a credible basis for claiming the reliability of such macro-remote and counterintuitive subjects as quantum physics or relativistic cosmology. Better grounded is Peacocke's appeal to inference to the best explanation (IBE).

I see the latter strategy as being as much exercised in theology as in science and I would answer Peacocke's question, 'Dare theology, by using IBE, enter the fray of contemporary intellectual exchange and stand up and survive in its own right?'[47] with a definite 'Yes'.[48] In contrast, he fears that current theological practice relies principally on an authoritative book, an authoritative community and an appeal to a priori truth. So fideistic an account would seem to be something of a

---

46. See W. van Huyssteen, *Duet or Duel?*, SCM Press, 1998, ch. 3.

47. Peacocke, 'Science and the Future of Theology', p. 130.

48. See Polkinghorne, *Faith, Science and Understanding*, chs. 1-3, and the approach of my Gifford Lectures, *Science and Christian Belief/The Faith of a Physicist*.

caricature. There then follows an interesting and detailed set of fourteen points that Peacocke believes to be essential for the new theology of the future. Many of these points refer, in fact, to issues that would be widely recognised as being on the contemporary theological agenda, whatever approach one took to the framing of theological discourse with science and with wider culture. They include such matters as the challenge presented by a monistic view of created reality; human beings seen as 'rising beasts' rather than 'fallen angels'; the possible theological implications of extraterrestrial life; God's relationship to time.

Yet there are other points that are intended to be a challenge to traditional Christian thinking. The assertion that 'the historical evidence for miracles is usually inadequate to testify to them' leads to the suggestion that theology should rid itself of dependence on ideas of the 'disruption of nature by God'.[49] Yet that Humean way of thinking about the miraculous is theologically very unsophisticated, for miracles are properly understood not as arbitrary divine acts, but as insights into a deeper rationality present in the divine relationship with creation than that which can be discerned through ordinary events.[50] Peacocke couples this discussion with a call to reassess the virginal conception and the bodily resurrection of Jesus.

To take another point, one may well agree that redressing the imbalance present in classical theology between divine transcendence and divine immanence is a desirable correction, but it is by no means evident that the best or only way to do

49. Peacocke, 'Science and the Future of Theology', p. 134.
50. J. C. Polkinghorne, *Science and Providence*, SPCK, 1989, ch. 4.

so is to embrace 'sacramental panentheism',[51] the notion that the world is in some way included within God.

I shall return to many of these matters as the argument of the book develops. The point at issue between the revisionary approach and the fourth possibility to be discussed below is not whether new scientific insights will have the possibility of influencing theology, or whether it is the case that each generation will have to make the fundamentals of the tradition its own in its own way, for many of us are revisionists in a more modest mode and theology has never been a purely static discipline. The essential issue is whether substantial new thinking in theology can satisfactorily be achieved largely in disconnection with past understanding. There is always the danger that the gusting of the *Zeitgeist* might wrongly be mistaken for the wind of the Holy Spirit.

(4) *Developmental.* This approach pictures the interaction between science and theology as a continuously unfolding exploration rather than a process of radical change. Although there are cousinly analogies between science and theology in their common search for truth (such as appeal to inference to the best explanation), there are also significant disanalogies. One of the most important of these is that theology does not share in the cumulative character that science displays. Scientific understanding of a well-winnowed and well-defined regime attains a stability that means that it will not require further revision or amplification unless the boundaries of that regime are crossed. Newtonian mechanics is still good enough

---

51. Peacocke, 'Science and the Future of Theology', p. 134; for a critique, see Polkinghorne, *Faith, Science and Understanding*, § 5.3; for an even stronger emphasis on sacramental panentheism, see C. Knight, *Wrestling with the Divine*, Fortress, 2001.

to get an explorer satellite to Mars, though it will need to be replaced by general relativity for motion in the neighbourhood of a black hole. Science conquers territory over which it then holds permanent sway. Consequently it is synchronic in its character, able to concentrate on the contemporary state of understanding.

Most humane disciplines, however, because of the complexity and subtlety of their subject matter, do not enjoy this kind of cumulative attainment. In consequence, their discourse cannot be confined simply to some present state of the art, but has to range over more than contemporary insights. There is no presumptive superiority of the twenty-first-century theologian over theologians from the fourth or sixteenth centuries, any more than there is a presumptive superiority of twenty-first-century art or literature over that of preceding centuries. Different generations gain different forms of spiritual insight and we have to be humble enough to be willing to apprentice ourselves to the past in a manner that is not necessary in science. As a result, theology is a diachronic discipline, for which dialogue across the centuries is an indispensable resource, not least as a means of release from the possible constrictions of a purely contemporary perspective. Karl Barth wrote that

> We cannot be in the church without taking responsibility for the theology of the past as much as for the theology of the present. Augustine, Thomas Aquinas, Luther, Schleiermacher and all the rest are not dead but living. They still speak and demand a hearing as living voices, as surely as we know that they and we belong together in the church.[52]

52. Quoted in A. E. McGrath, *Nature 1*, T&T Clark, 2001, p. xv.

To acknowledge the truth of this is not to put oneself in un-thinking thrall to the past, as Peacocke seems to fear, but to make the fullest use of the resources available for theological progress. Of course, in the twenty-first century, as in every century, people have to appropriate these resources in their own way, and by doing so the insights from the past will be modified and qualified. Yet the change will be evolutionary rather than revolutionary, so that the adjective 'developmental' is the right one to apply to the process. After all, this is just what one might expect to be the case as a result of the continuing work of the Holy Spirit, guiding and leading the Church into further truth (John 15:26; 16:12–14). It would be very surprising if what Christian belief is really all about only came to be realised in our time, just as it would also be very surprising if previous generations had already so perfectly comprehended Christian truth for there to be nothing left for us to do but passively accept their conclusions.

It is the developmental approach that has been the stance that I have sought to adopt in my contributions to the dialogue between science and theology. In the introduction to my Gifford Lectures, I stated that I sought a 'basis for Christian belief that is certainly revised in the light of our twentieth-century insights but which is recognisably contained within an envelope of understanding in continuity with the developing doctrine of the Church throughout the centuries'.[53] I described the method to be employed as 'bottom-up thinking', an abductive strategy that seeks to take the record of experience as the basis for the search for understanding. Such a quest for motivated belief seems to me to be of a kind that

---

53. Polkinghorne, *Science and Christian Belief/The Faith of a Physicist*, p. 8.

responds to Peacocke's challenge that theology should 'enter the fray of contemporary intellectual exchange'. It certainly has the character of inference to the best explanation. I sought in the course of those Gifford Lectures to consider motivations for Nicene Christian belief, stating my conclusion that 'the Nicene Creed provides us with the outline of a rationally defensible theology which can be embraced with integrity as much today as when it was first formulated in the fourth century'.[54] One of the reasons why that claim can be made is that the condensed character of the Creed means that it indicates the heads of an adequate Christian discourse, without prescribing all the details of how that discourse must be conducted. Much the same can be said about the Christological Definition of Chalcedon, which stakes out a ground within which it believes the understanding of the Church's experience of Christ must be located, without attempting to mark a single point from which to deviate would be seriously heretical. Orthodoxy is not inflexibility.[55]

A comparison between the way in which these last three approaches function within Christian thinking can be made by seeing briefly how they influence consideration of the question of the virginal conception of Christ. Barbour does not discuss the matter at all, but Peacocke and I have both paid attention to the issue.[56] We agree that the evidential testimony

54. Ibid.

55. There is a connection between these thoughts and John Henry Newman's concept of the development of doctrine, but I would always want to find a scriptural seed from which the subsequent elaboration of understanding had grown. In my view, the Marian dogmas fail this test.

56. Peacocke, *Theology for a Scientific Age*, pp. 275-79; Polkinghorne, *Science and Christian Belief*, pp. 143-45; *Scientists as Theologians*, pp. 78-80.

offered in the New Testament is far weaker than the corresponding testimony to the resurrection, so that for both of us the central issue to be resolved is that of the relevance of a criterion of theological coherence to the assessment of the claim. Peacocke begins his consideration with a long discussion of where Jesus' Y chromosome could have come from. (As a woman, Mary would only have had X chromosomes.) This is really a restatement in modern form of the ancient understanding that if the virginal conception actually took place its character was miraculous rather than natural. Peacocke then questions whether this exceptional status would not deprive Jesus of identification with the rest of us who have been naturally conceived. This is a very serious point, for we both agree that Christian theology demands that Jesus is to be understood as being fully and truly human. I cannot see why this is not fulfilled by Jesus having a human genome, whatever its origin might have been. To suppose the contrary would be an instance of the so-called genetic fallacy, that nature is determined by origin. In any case, the total humanity of Jesus is most clearly demonstrated by the way in which he shared absolutely in human death, even to the point of reluctance at its approach (Gethsemane: Mark 14:32–42 etc.) and to a feeling of God-forsakenness in the darkness of the end (Mark 15:34; Matthew 27:46). Of course, Jesus' death was followed by the uniqueness of Jesus' resurrection on the third day, but it is important that this was not immediate, for between Good Friday and Easter is the silent grave and real death of Holy Saturday.[57] Moreover, Jesus' resurrection within history is to be understood as the anticipation and guarantee of what awaits

57. See A. K. Lewis, *Between Cross and Resurrection*, Eerdmans, 2001.

all humanity beyond history: 'as all die in Adam, so all will be made alive in Christ' (1 Corinthians 15:22).

For me the central issue relating to the virginal conception is different, focusing on whether this story, whose symbolic significance is clear enough (the combination of divine initiative and human participation in the coming-to-be of Jesus), is also properly required theologically to be an *enacted* story. Since I see the whole force of Christian incarnational belief as deriving from the fusion of the power of symbol with the power of actual history, I am prepared to believe that the virginal conception actually took place, as the inception of the salvific episode of true divine sharing in the life of humanity. The difference between the revisionary and the developmental approaches may often be found to lie in the differing ways in which they estimate the relative importance of scientific expectation and theological interpretation in weighing the probability of specific beliefs. I believe that Christian salvific symbols are never merely free-floating, but always anchored in actual occurrence, a principle that one might call sacramental.

Lastly, I realise that it may seem strange that my four typical thinkers have all been people whose background is scientific rather than theological. Does this just illustrate the parochiality of the scientist-theologian? I do not think so. While there are certainly theologians who take an interest in the dialogue with science,[58] their writing is often concerned with largely methodological issues and shows little sustained

---

58. The contributions of Wolfhart Pannenberg and Thomas Torrance are discussed in Polkinghorne, *Faith, Science and Understanding*, pp. 156–85.

engagement with the content of the natural sciences. I believe that such a content-based consideration is indispensable. At the same time, I am sure that the insights and questions that mainstream theologians could bring to the conversation would be of great value. I hope that in the future there will be a fifth approach, of a kind that one might label 'Systematic'.

CHAPTER TWO

The Role of Scripture

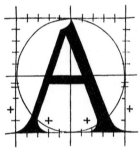LL faith traditions have a body of authoritative writings that are normative for that tradition and which act as scripture within it. For the Christian, of course, this role is played by the Bible, the combination of the Jewish Hebrew Bible, or *Tanach*, together with the Greek New Testament, originating in the post-resurrection Church. This collection of writings contains very diverse material deriving from a period spanning at least a thousand years. Almost two thousand years separate us from the composition of the last of these writings. What then could be their legitimate role for a believer at the beginning of the twenty-first century?[1] Is it that of the unquestionably authoritative book that we saw that Arthur Peacocke feared, or is it

1. A related, but somewhat differently oriented, approach to the issues of scripture is given in J. C. Polkinghorne, *Reason and Reality*, SPCK/Trinity Press International, 1991, ch. 5.

something altogether more subtle and more interesting than that?

The first thing to say is that virtually all Christian theologians acknowledge that there is a special role for scripture, and that opinion is certainly held by the scientist-theologians. Ian Barbour refers to the Bible from time to time and his thought is clearly influenced by general biblical ideas, even though his writings contain comparatively little extended engagement with the detailed text of scripture.[2] Barbour favours the approach of narrative theology, with its emphasis on the illuminating power of story, but this does not cause him to neglect the importance of historicity: 'If no Exodus took place, and if Christ did not go willingly to death, the power of the stories would be undermined'.[3] It is difficult for any theologian not to be somewhat eclectic in the use made of the wealth of imagery and concepts that the Bible provides. Barbour's choice is understandably influenced by his metaphysical predilections, and he tells us that 'it is in the biblical idea of *the Spirit* that we find the closest fit to the process understanding of God's presence in the world'.[4]

Peacocke, on the other hand, devotes much more attention to biblical detail, particularly in relation to the New Testament witness to the life of Jesus of Nazareth.[5] In assessing this material, he confesses that he is inclined to adopt 'a

2. An exception is I. G. Barbour, *Religion in an Age of Science*, SCM Press, 1990, pp. 204–209.

3. Ibid., p. 72.

4. I. G. Barbour, *When Science Meets Religion*, HarperCollins, 2000, p. 176, his italics.

5. A. R. Peacocke, *Theology for a Scientific Age*, enlarged edition, SCM Press, 1993, ch. 13.

more "trusting" attitude to the scriptures'[6] than is the case with many professional New Testament scholars. I certainly agree with Peacocke in taking this stance, and I am encouraged in doing so by the way in which, for example, the gospel writers record words that must have given as much difficulty to them as they do to us (such as the cry of dereliction from the cross (Mark 15:34; Matthew 27:46), or Mark and Luke's version of Jesus' reply to the rich young man, 'Why do you call me good?' (Mark 10:18; Luke 18:19)). One gets the distinct impression that the writers are trying to tell it how it was, within the limits of their knowledge and in accordance with the historical conventions of their day.

The essence of good scholarship is not a detached scepticism but a willingness to approach the material being considered on the terms that are appropriate to its intrinsic nature. It is not to some abstractly conceived notion of prior rationality that we must strive to conform our thinking, but to the character of what it is that we are seeking to understand.[7] A physicist who attempts to approach an intrinsically quantum phenomenon, such as the photoelectric effect, but is only prepared to do so in accordance with the thought patterns of classical physics, will be condemned to permanent bafflement. The central claim of the writers of the New Testament is that they are responding to a divine initiative and to a revelatory event that are without precedent, even to the extent that therein is involved the resurrection of a man from

6. Ibid., p. 262.

7. This is a persistent theme in the writing of T. F. Torrance: see *Theological Science*, Oxford University Press, 1969; *Reality and Scientific Theology*, Scottish Academic Press, 1985. See also J. C. Polkinghorne, *Faith, Science and Understanding*, SPCK/Yale University Press, 2000, pp. 173–85.

the dead to an unending new life of glory. Those who refuse altogether to countenance the possibility that this might be so are condemning themselves to a Humean dismissal of anything that suggests a happening out of the ordinary. They also force themselve to take up a position that makes the phenomenal and persistent influence of Jesus, and indeed the very existence of the New Testament writings themselves, inexplicable facts. It seems quite inadequate to consider them as stemming simply from the case of yet another good man sadly done to death by the intransigent opposition of the forces of the worldly-wise. *God doesn't interviee in peoples lives - no miracles*

scientific view ← A flatly naturalistic account of the New Testament, predicated on the supposition that what usually happens is what always happens, faces insuperable difficulties because it has prejudged exactly the issues that are at stake in the investigation. That scrupulous and extremely helpful historical writer, E. P. Sanders, ends his careful discussion of the life of Jesus by saying, 'What is unquestionably unique about Jesus is the result of his life and work. They culminated in the resurrection and in the foundation of a movement that endured'. Despite Sanders' use of that word 'resurrection', his historian's caution does not allow him to give it its proper force. He goes on to say, 'I have no special explanation or rationalisation of the resurrection experiences of the disciples'.[8] Someone whose writing is bound by secular conventions can say little more. An adequate consideration of the resurrection demands the resources that a theological approach can alone supply.

The proper fulfilment of the role of scripture is hampered within the academic world by the convention that so

8. E. P. Sanders, *Jesus and Judaism*, SCM Press, 1985, p. 320.

often separates biblical studies from systematic theology. In its search for truth, theology cannot do without the aid of careful scholarly investigation into the nature of its central texts, but that analysis cannot properly be conducted on the basis of imposing an atheological grid of interpretation on the material to which those texts refer. It must be admitted that this acknowledgement that theology and biblical studies need each other does introduce a measure of circularity into the resulting discussion, but no more than is the case in the interaction between theory and experiment in natural science.[9] Neither theology nor science has access to 'plain facts' that are interesting and significant without the need for any further interpretation. Neither subject can escape the necessities of the hermeneutic circle, encountered as those who pursue their discipline find that though they have to understand in order to believe, they also have to believe in order to understand. What must be done is to tighten the circle sufficiently so that it is seen to be benign and not vicious.[10]

The indispensability of the role of scripture in the task of theology is clearly expressed by Peacocke when he writes, 'The faith of the Christian church derives from its experience, the principal resource and source for which are the archetypal and seminal experiences and encounters with God recorded in its scriptures'.[11] He describes the Bible as being 'a unique and irreplaceable resource'.[12] I agree very strongly with that assessment. I would also agree with Peacocke's criticism of those

9. See J. C. Polkinghorne, *Beyond Science*, Cambridge University Press, 1996, ch. 2.

10. J. C. Polkinghorne, *Science and Christian Belief/The Faith of a Physicist*, SPCK/Princeton University Press, 1994, p. 32.

11. Peacocke, *Theology for a Scientific Age*, p. 94.

12. Ibid., p. 339.

who treat the Bible as 'a kind of oracle, as if quotations from past authorities could settle questions in our times',[13] if this is to be understood both as a repudiation of a 'soundbite' approach based on appeal to isolated 'proof texts', and also as the recognition that there are many particular issues faced by Christians in our generation that do not have clear precedents in the experience of previous generations. Many of these latter questions arise from the advance of science, illustrated, for example, by the perplexities generated by the rapid rate at which techniques for genetic engineering are currently being developed.

What then is the proper way in which to use scripture? The answer is necessarily complex, since there is not a single kind of usage involved.

First, there is an indispensable *evidential role* for scripture. From extra-biblical sources alone, we could glean very little about either the history of Israel or the person of Jesus of Nazareth. The claim that those people and that person represent uniquely important loci in which the divine nature and the divine purpose have been most clearly revealed depends both for its justification and for its content on our being able to gain reliable knowledge of that history and that life. To someone with a bottom-up habit of thought, this evidential role of the Bible will be of great importance. I devoted a significant fraction of my Gifford Lectures to seeking to evaluate what we can properly believe we know about the life, death and resurrection of Jesus Christ.[14] Of course, the way in which the evidence is weighed will also depend upon how persua-

13. A. R. Peacocke, 'Science and the Future of Theology: Critical Issues', *Zygon* 35 (2000), p. 130.

14. Polkinghorne, *Science and Christian Belief/The Faith of a Physicist*, chs. 5–7.

sive one finds the theological understanding offered as the interpretation of the results obtained. There is no escape from the hermeneutic circle. Nor is there an escape from facing the question of the possibility of unique, and uniquely significant, events. I do not want on this occasion to explore in any detail the evidential use of scripture, but I shall simply content myself with registering my considered opinion that the rational use of the Bible in this mode provides a foundation for Christian belief that can be defended with intellectual care and scrupulosity. The nearest scientific analogy is not with the repeatable and controllable investigations of experimental science but with the unique and fragmented records that are the basis for the insights of historico-observational sciences such as evolutionary biology and physical cosmology. In both science and theology it is necessary to have recourse to creative interpretation based on making overall sense of the data available. Finding this understanding may be a lengthy process. It does not worry me, therefore, that one does not find in the New Testament itself a fully articulated expression of Trinitarian doctrine (despite those remarkable verses, Matthew 28:19 and 2 Corinthians 13:14). Rather, one finds the raw materials of encounter with the reality of God, Christ and the Spirit that eventually led the Church to a Trinitarian conclusion.

Second, let me turn to the *spiritual use* of scripture. Here we do not sit in judgement on the character and historicity of the text, but we allow that text to sit in judgement on us, as we seek to make it the inspiration and guide of a pilgrim life lived in the presence of God. One aspect of this usage is the way in which scripture is employed in the liturgy. As an Anglican priest fulfilling my obligation to say the Daily Office, I read through the whole of the New Testament every year and through long tracts of the Old Testament also. Most importantly, I also pray through the whole psalter 'in course' (that is, in sequence) every ten weeks.

The Psalms have a very special place in the spiritual use of scripture, a role that they have played throughout a period that stretches over two and a half thousand years, from the worship of the Second Temple in Jerusalem, through the centuries of Christian monastic practice (where many religious communities pray the entire psalter every week), to their prayerful use in common worship at the present day. In miniature, they encapsulate both the power and the problematic of scripture.

On the one hand, the Psalms include many fine passages expressing human response to the majesty and glory of God: 'The Lord is King! Let the earth rejoice; / let the many coastlands be glad!' (Psalm 97:1). On the other hand, they include many passages that express extreme human feelings of merciless vindictiveness. Psalm 137, which starts so beautifully with lament over the sadness of exile in Babylon: 'By the rivers of Babylon there we sat down and there we wept / when we remembered Zion' (v. 1), ends with a terrible cry against the captors: 'Happy shall they be who take your little ones / and dash them against the rock!' (v. 9). What are Christians, or Jews for that matter, doing if they take these words on their lips in worship? Certainly they cannot be uttered uncritically, as if they were expressions of what we understand to be the way that the servants of God should think and behave.

The significance of unedifying passages in the Bible is one to which I shall return, but it is already clear that scripture does not portray a sanitised picture of human nature or human history. The Bible does not give us an idealised account in which only the good and the beautiful are recorded. On the contrary, the deep ambiguities of the human condition, and the power of human sinfulness, are frankly displayed in its pages. While the violent passages in the Psalms rightly disturb us, they also remind us of what still lurks in the distorted depths of the imperfect human heart. A great part of the power of the Psalms resides precisely in their honesty. They embrace a much wider and more troubling range of human spiritual experience than can be found, for example, between the covers of a hymn book. The psalmists speak with great frankness to God, often asking why their distress has been overlooked and their needs forgotten. Psalm 44 even calls on God to wake up: 'Rouse yourself! Why do you sleep, O Lord?' (v. 23).

Especially powerful and helpful are the Psalms of lament, which begin with discontent but end, complaint notwithstanding, in an affirmation of trust in the God of Israel— a trust that is surely all the more profound because it has emerged from the honest protest that preceded it. Psalm 13 is a model in this respect, with its opening line, 'How long, O Lord? Will you forget me for ever?' and its closing verse, 'I will sing to the Lord / because he has dealt bountifully with me'. The Christian will call to mind Gethsemane, where Jesus was 'deeply moved, even to death' and where his prayer began, 'Father, for you all things are possible, remove this cup from me' but ended 'yet not what I want but what you want' (Mark 14:33-42). Jesus' acknowledgement of his great distress takes place in the presence of God. Even more profoundly may we say that of the cry of dereliction from the cross, 'My God, my God, why have you forsaken me?' (Mark 15:34), for the words that come from out of the darkness of Calvary are still addressed to God. Although these words are the first line of Psalm 22, that cry of Jesus in the context of his passion goes far beyond the limits of liturgical use in its plunge into the abyss of the sense of God-forsakenness. The distance is as great as that which separates the twisted and scarred figure of Grüne-wald's Isenheim altarpiece from the tolerable proprieties of a conventional crucifix.

The spiritual use of the Bible is a resource that can be employed in a variety of ways. One of the most valuable is the method that monastic piety calls *lectio divina*, in which a short passage, or perhaps just a single verse, is meditated upon reit-eratively, allowing its truth to dissolve in the mind and to be absorbed into the heart of the reader. Such practice is very far from the crisp analytic assessment that the scholar may make of the genre and probable provenance of the passage. Both kinds of engagement have their place in the multi-roled usages of scripture, for both are concerned with its truth-bearing character, though the aspects of truth thus encountered are met with at very different levels of human experience.

There is a further role for scripture that is particularly important for theology but whose nature is somewhat elusive and difficult to define. One may call it the *contextual role*, for it is concerned with structuring the setting within which theo-logical reflection takes place. We are familiar with the notion of contextual theologies that derive insight from a particular perspective, such as concern for feminist issues or with politi-cal liberation, but the basic context of all Christian theologi-cal thinking is that provided by scripture itself. Immersion in the Bible provides a panoply of images, and a perspective on reality, that together constitute a fertile ambience for think-ing about God. I do not mean by this simply an engagement with those great revelatory insights that centre on such critical episodes as the Exodus from Egypt, the Babylonian Exile, the life, death and resurrection of Jesus Christ, and the pouring out of the Spirit at Pentecost. Although we depend upon scripture for our knowledge of these singular times of special revelation (*kairoi*, in the Greek idiom), there is a great deal in the Bible that does not directly relate to them. The whole range of biblical material helps to generate the atmosphere that we breathe, so to speak, as we think about matters theological. Jürgen Moltmann seems to me to be saying something like this when he writes,

> Every Christian theology, however conditioned it is by context, *kairos* and culture, follows and interprets the text of the biblical writings. So it is important for everyone *who exists within the orbit where the Bible is interpreted*, wherever they live, whenever they live, and whoever they may be. For it is the text which determines what for it is the context.[15]

It is comparatively easy for those of us who seek to operate 'within the orbit where the Bible is interpreted' to recognise each other, even if sometimes we find the other person saying something very different from what we ourselves think. A respect for the testimony of scripture can be expressed in a variety of forms, and it is certainly not confined to those who make ready and repeated recourse to the formula 'The Bible says . . .'. Indeed, respect for the integrity of scripture will sometimes include the recognition that it is by no means always clear what it is that the Bible is actually saying on a particular topic, either theological or ethical. Scripture is very

15. J. Moltmann, *Experiences in Theology*, SCM Press, 2000, p. 60, my italics.

far from being the handy textbook in which we can conveniently look up ready-made answers. This contextual role of the Bible raises in an acute form the hermeneutical question of how scripture is properly to be interpreted. One may identify a number of controlling factors.

The first is the simple necessity of identifying the *genre* of what we are reading. The Bible is often rightly said not to be a book but a library. It contains a great variety of different kinds of writing: poetry and prose, history and story, letters, laws, and so on. Very great mistakes can be made, and much disrespect shown to scripture, if a reader carelessly confuses one genre with another. Those who attempt to read Genesis 1 and 2 as if these chapters were divinely dictated scientific texts, kindly provided by God to save us the trouble of attempting to read the book of nature for ourselves, are committing just such an act of literary violence. They also put themselves in peril of missing the true theological point of the text, with its eightfold reiteration of the message that nothing exists except through the creative will and effectual utterance of God ('And God said "Let there be . . ." '). Those who disdain a scholarly engagement with the same text will also miss the fact that, though the accounts are clearly influenced to a degree by neighbouring Near Eastern cosmogonies, they differ in a most marked and important way from those other creation stories. It is deeply impressive that tales of conflict among the gods, with Marduk fighting Tiamath and slicing her dead body in half from which to form the earth and sky, are replaced by a sober account in which the one true God alone is the Creator, bringing creation into being by the power of the divine word. Equally significant is the insight that human beings are not destined to be the slaves of the gods (as in the Babylonian epic, *Enuma Elish*), but are created in the image of God and given a blessing so that they may fulfil the command, 'Be fruitful and multiply and fill the earth and subdue it' (Genesis 1:28).

A genre of scripture that has been particularly valued by those in the science and theology community has been that of the wisdom writings. They are at least as important a source for a biblical attitude to nature as are the first two chapters of Genesis. The sages take a cool look at the world, seeing things as they are and expressing themselves in uncompromising terms: 'Like a gold ring in a pig's snout is a beautiful woman without good sense' (Proverbs 11:22). They know that 'the fear of the Lord is the beginning of wisdom' (Proverbs 9:10), but their attention is mostly centred on the works of the Lord in creation, rather than on salvation history. Hence the astonishing answer that is given to Job out of the whirlwind, simply to look at the wonders of the natural world around him (Job 38-41). The wisdom writers are the natural theologians of the Old Testament. Perhaps the nearest we get to their attitude in the New Testament is in something like Acts' account of Paul's address to the secular enquirers of the Areopagus (Acts 17:22-34), with its appeal to the resources of general culture. Natural theology is a limited exercise, but one that provides a meeting place where different traditions can encounter each other. We see this happening when the compiler of Proverbs borrows whole chapters more or less verbatim from Egyptian wisdom sources, a strategy one could not imagine being used anywhere else in the Hebrew Bible.

The second factor controlling biblical interpretation is the need to recognise, as Christians at their best have always done, that scripture is *divinely inspired but not divinely dictated*. These human writings bear witness to timeless truths, but they do so in the thought forms and from the cultural milieu of their writers. It is not surprising, therefore, that we find attitudes expressed in the Bible that today we neither can nor should agree with. These include an unquestioned patriarchal governance of the family, with a consequently depressed status for women, and an unhesitating acceptance of the institution of slavery in society, though this was significantly qualified within Israel by the limitation to seven years of servitude (Exodus 21:2-6). Those who attribute no abiding significance to these timebound attitudes are recognising that the canon of scripture is not of uniform authority. While some passages undoubtedly have enduring significance, others prove to be dispensable in the light of developing insight. We see this happening within the span of the Bible itself, when the early Church released Christians from obedience to Hebrew dietary law and from the *Torah* command to circumcise all males in the household of God (Acts 15:12-21). In actual fact, whatever their formal position about the authority of the Bible, absolutely no one treats the whole of scripture as being equally significant and unexceptionally binding in every respect. (Think of the biblical disapproval of usury.) The task of sifting and testing scripture within the community of the Church is one that may readily be understood as being conducted under the guidance of the Holy Spirit. The Spirit's continuing influence was surely at work when, after eighteen centuries, the Christian conscience came to recognise that slavery was repugnant to it. One may say the same about the somewhat later, and theologically more controverted, questioning of whether the many New Testament passages that speak of fearful and fiery judgement are rightly interpreted as implying that the God and Father of our Lord Jesus Christ requires the punishment of unending torture for those who have not committed themselves to some kind of Christian orthodoxy in this life. To deny the kind of concept of hell that we find in Revelation 20, and in Dante, is not to deny the seriousness and life-denying character of persistent refusal of the divine mercy, but rather it is radically to modify and revalue the images through which that seriousness is expressed.[16]

Similar issues are raised, even more intensely, by asking how we should understand the many *unedifying incidents* that are to be found in the Bible. After reading a lesson like Numbers 15:32–36, the story of the man caught gathering sticks on the Sabbath who is taken outside the camp and stoned to death 'just as the Lord commanded Moses', the liturgical conclusion 'This is the word of the Lord' may well seem to stick in the throat of the lesson reader. In the Anglican lectionary, the reading retailing the storming of Jericho by the Israelite army under Joshua ends at chapter 6, verse 20, thereby omitting what is really the concluding verse of the biblical story: 'Then they devoted to destruction by the edge of the sword all in the city, both men and women, young and old, oxen, sheep and donkeys'. While this act of pious dishonesty on the part of the lectionary compilers is entirely understandable, it is also an act of violation imposed on the integrity of the chilling text of scripture. Even the lectionary, however, cannot dodge the difficult story that follows, of the sin of Achan and of his being stoned to death, together with all his innocent family. What makes these ancient and violent tales particularly difficult is that the events they record are presented as being direct ful-

---

16. See J. C. Polkinghorne, *The God of Hope and the End of the World*, SPCK/Yale University Press, 2002, pp. 136–38.

filments of the commands of God. Among the most disturbing in this respect is the divine command to Saul to commit genocide against the Amalekites (1 Samuel 15:3). Unedifying events of this kind are predominantly located in the Old Testament, but the drastic and summary punishment meted out to Ananias and Sapphira for an act of deceit (Acts 5:1-11) is a morally perplexing incident in the New Testament. It seems to me that the Christian reader has to react to such stories not by unquestioning acceptance, but by a perplexed reserve or by radical revaluation.

In the interpretation of the Hebrew Bible, the exegete has often to wrestle with the combination of an unsophisticatedly anthropomorphic style of expression together with profound underlying theological themes and symbols. A striking example is the story of the binding of Isaac (Genesis 22). At one level this is a disturbing tale of a cruel test imposed upon Abraham and his hapless son. At another level it carries the message that the religious custom of child sacrifice, though widespread in the ancient world, is not pleasing to God. At a deeper level still, the story has inspired continuing Christian analogical reflection on the significance of Christ's sacrificial death on the cross. Another example of the strange way in which scripture can interleave the limited and culturally particular with the open and theologically profound is to be found in the book of Job. The deep explorations of the problem of suffering that make up the poetic centre of the book are prefaced by the prose tale of an apparently amoral wager in the heavenly court about Job's constancy under misfortune, and concluded by a folk-tale ending in which Job is given twice as much again to 'recompense' him for his previous losses.

These considerations lead me to conclude that it is impossible to read scripture honestly and without an undue degree of mutilation unless one is prepared to recognise its *developmental character.* That is to say, there is contained within the Bible the record of a long process of spiritual exploration and encounter with the reality of God, in the course of which early insights are refined, revalued and expanded. Of course, for the Christian this happens most significantly when the coming of Jesus Christ leads the Church to a rereading of the Hebrew Bible in the light of the events of the incarnation, resulting in the recognition that God's Chosen and Anointed One is not a Davidic military deliverer but a crucified Messiah.

I am sympathetic to the canonical insistence that theologically one must seek to read scripture as a whole, neither reducing it to a sanitised collection of easily acceptable passages nor decomposing it into a multitude of form-critically-dissected fragments.[17] Part of our engagement with scripture is to allow its strangeness to challenge our preconceptions. Yet this is only possible if we are also prepared to be sufficiently aware of the historical frame of the Bible as to be able frankly to acknowledge the developmental character of its material.

The Fathers had a different way of dealing with those parts of scripture that seemed contrary to Christian understanding when they were read straightforwardly at their apparent face value. Patristic thought employed a variety of schemes, identifying a multiplicity of levels at which the Bible might properly be interpreted. There were the literal, the moral, the symbolical and the spiritual senses to be consid-

17. See B. Childs, *Introduction to the Old Testament as Scripture*, SCM Press, 1979; *The New Testament as Canon*, SCM Press, 1984.

ered, and the methods of exegesis used might rely on appeal to allegory or to the typological notion that some Old Testament events were foreshadowings of what was to be fully expressed in New Testament happenings. Today we may well not want to follow any of these schemes in their detail, but they point us to a very important factor in the interpretation of scripture, namely the essential need to acknowledge the *polysemous character* of the text. The biblical writings are diminished if they are restricted to carrying a single sense. We have already seen something of the multiplicity of scriptural meaning in our brief consideration of the story of the binding of Isaac. Another aspect of polysemy is represented by the multi-stranded writings of the Pentateuch, resulting from different generations reworking archetypal material in the course of further exploration of the meanings to be found there.

There is surely importance to be attached to the original meaning of a piece of writing as intended by its author, if that can be recovered with a degree of plausibility by means of historical and cultural considerations. However I could not agree with certain theories of the role of scripture which assert that this is necessarily the sole or the determinative meaning. Who could doubt that the words 'You are a priest for ever according to the order of Melchizedek' (Psalm 110:4) were originally intended as the affirmation, or even the flattery, of a King of Judah at his coronation? Yet, in the hands of the writer to the Hebrews, the verse legitimately becomes the ground for an extended meditation on the royal priesthood of the exalted Christ (Hebrews 5:5–7:28). One of the most impressive examples of the creative development of meaning is afforded by the history of the interpretation of the Song of Songs. The book's origin obviously lies in its being a powerful expression of the erotic love between a man and a woman. At first sight, it might seem highly ironic that, in the hands of medieval ascetics such as Bernard of Clairvaux, Canticles became the source of many sermons and meditations on the spiritual love between Christ and the individual human soul. (There are some similar thoughts, though less exuberantly expressed, in the Pauline writings, e.g. Ephesians 5:32–33). Nevertheless, such is the profound power of symbolic imagery that this appropriation of human sexuality may rightly be accepted as the expression of a deep analogy, one that is articulated in an entirely different way in Bernini's famous sculpture of St Theresa of Avila in religious ecstasy as an angel stands ready to pierce her heart with the arrow he is holding.

Belief in the continuing work of the Holy Spirit undergirds a theological understanding of the fruitfully open meaning of scripture, for it allows for a process by which the unfolding of divine truth has occurred in the past and can continue to occur into the future. Christian reliance on the Bible implies reliance on the activity of the Spirit in the inspiration of the writings, in the assembly of the canon (both of the New Testament and of the Hebrew *Tanach*), and in the Church's unending exploration of biblical insight. There are obvious dangers involved in such an open view, for it can lead to the wilful or fantastic manipulation of scripture, but amongst all who 'exist within the orbit where the Bible is interpreted', one can detect a degree of control that mediates between a narrow concept of original authorial intention and a loose and individualistic reader response, much as the philosophical attitude of critical realism mediates between a modernist claim to the possession of clear and certain ideas and an extreme postmodernist abandonment to rampant relativism.[18] Pursuing this analogy with the philosophy of science leads to the conclusion that just as individual scientists are the indispensable originators of ideas, yet have to submit their proposals to the judgement of the competent community of their peers,[19] so what is often thought of as a Protestant emphasis on the individual believer's right to read and interpret scripture has to be qualified by what is often thought of as a Catholic emphasis on the sifting and receiving role played by the whole Christian community.

Not all of scripture, however, yields a richness of multi-levelled theological interpretation. Much of the Hebrew Bible is simply historical in form and I believe—contrary to much contemporary scholarly fashion—largely in content also. By no means all of this material relates to what one might properly call salvation history. The absorbing account of palace intrigue taking place around the throne of an ageing king that constitutes 'the succession narrative' (2 Samuel 17–1 Kings 2), telling how eventually the young Solomon came to follow his father David as king of the united kingdom of Israel and Judah, is fascinating reading. It reveals a great deal about human nature's duplicity and ambiguity, but why does an Anglican priest have to read it every year? I think that the answer may lie in Romans 11 (vv. 17–24), where Paul speaks of the Christian Church as a wild olive branch that has been grafted onto the olive tree of Israel. In recent years, many Christians have recovered a fuller sense of the Jewish rooting of our faith

18. See for example, J. C. Polkinghorne, *Belief in God in an Age of Science*, Yale University Press, 1998, ch. 5.
19. M. Polanyi, *Personal Knowledge*, Routledge and Kegan Paul, 1958.

and of the debts and obligations that this means we owe to Judaism. By becoming Christians, we have acquired a new set of ancestors, and that is why the whole history of Israel is important for us, for by grafting and adoption it has become part of our history also.

What then shall we make of the role of prophecy within that history? I have already noted how the Christian interpretation of Messianic prophecy has introduced an intratextual criterion into the understanding of scripture. It seems to me that it is consonance of understanding, rather than confirmation of prediction, that is involved. I think one can see that this is true even in a writing substantially influenced by what might seem to be a more conventional theme of straightforward fulfilment. I have in mind the Gospel of Matthew. It contains significant emphasis on happenings in the life of Jesus that are to be seen as corresponding to what had been spoken of beforehand by the prophets. But the truth is that no one simply searching the Old Testament without preconceptions would come up with so oddly assorted a collection of texts, interpreted in the way they are by Matthew. For example, no one looking for insight into what might be expected of the Messiah would take the reference to the Exodus in Hosea's phrase 'Out of Egypt I called my son' (11:1) and suppose that it also implied that God's Anointed must sojourn in that country. Despite Matthew's tacit concern with the theme of Jesus as a second Moses, I think it likely that he cites this verse because he has an independent reason for believing that was what happened to Jesus in infancy. In consequence, I am not willing to dismiss the possibility that the Holy Family did indeed flee into that country, precisely because it seems gratuitous to suppose that the story was invented to fit in with an isolated text from Hosea, arbitrarily selected from the wealth of possibilities present in Hebrew prophetic utterance.

An even clearer case of intratextual flexibility is Matthew's citation (3:3) of Isaiah 40:3 applied to John the Baptist. In the original Hebrew 'in the wilderness' relates to the way of the Lord, but Matthew repunctuates the verse to make it correspond to the undoubted historic fact of the Baptist's appearance in the wilderness.

Jürgen Moltmann says that 'God promises but does not prophesy',[20] meaning that God does not use the prophets to provide a detailed preview of what must inevitably come to pass. Rather, God proclaims the divine attitude of covenantal love within which future events will happen. An important factor in how one interprets scripture will be how one conceives of *God's relationship to time and to historical process.* Modern science discerns a world that is dynamically open and evolving and not statically mechanical and deterministic. The theological counterpart to these ideas is the conception of cosmic history as an unfolding creative improvisation rather than the performance of a divinely pre-ordained score. The scientist-theologians believe that, as part of the divine kenosis involved in the act of bringing into being the created other, allowed to be itself and to make itself, God has freely embraced temporality in addition to divine eternity, even to the point that, in a creation that is a world of true becoming, God does not yet know the unformed future, simply because it is not yet there to be known.[21] Although such a view runs counter to the thinking of classical theology, it certainly seems consonant

20. Moltmann, *Experiences in Theology,* p. 96.
21. See J. C. Polkinghorne, *Scientists as Theologians,* SPCK, 1996, p. 41; J. C. Polkinghorne (ed.), *The Work of Love,* SPCK/Eerdmans, 2001.

with the God of the Bible, who interacts so intimately with the history of Israel and who totally accepts the experience of temporality in the incarnation of the Son. In its anthropomorphic way, the Hebrew Bible can even speak of God as changing the divine mind in response to events, as when Hezekiah is first told to prepare for death but then is granted fifteen more years of life (2 Kings 20:1, 6). This is a theme to which I shall return in Chapter 4.

The God who is detached from time, able to see the whole of history at once, might indeed be thought to be self-revealing by the communication of timeless truths. The corresponding view of scripture might tend to be that of the determinatively authoritative book. The God whose purposes are being worked out through the unfolding improvisations of open historical process will be known in a correspondingly more open and developmental way. In this case, the resulting view of scripture will be something like that which I have been trying to outline.

The cumulative effect of the considerations I have discussed is to encourage the belief that the proper role of scripture is as the record of God-given experience and insight, which has a continuing life of its own under the guidance of the Holy Spirit. If it is to be properly understood, this will call for a subtly *flexible hermeneutic*, not meaning by that that anything goes, but rather that scripture is not to be confined in the straitjacket of a single interpretation. I feel confirmed in this view when I consider the intratextual use of scripture exhibited by the New Testament writers when they have recourse to the heritage of the Hebrew Bible, which for them, of course, was exactly what 'scripture' meant. I would like to close this chapter by a brief consideration of typical ways in which Paul appeals to what we would call the Old Testament in the course of his most systematically doctrinal writing, the Epistle to the Romans. In its opening verses, the letter immediately locates the gospel that Paul is proclaiming within the scriptural context of what God 'promised beforehand through his prophets in the holy scriptures' (Romans 1:2). We shall see that Paul's understanding of the role of scripture involved a considerable degree both of respect for his sources and also of creative freedom in the way that they were used.

In principle, Paul had available both the Hebrew text of the *Tanach* and also the Greek translation of the Septuagint. Sometimes he seems to quote from one, sometimes from the other, and sometimes from neither. Since Paul may well often have been quoting from memory, it is hard to be sure what significance, if any, to attribute to this. It is clear, however, that Paul felt free to adapt the form of the Greek that he used in his quotations in Romans in order to suit his theological purposes. In Romans 3:10, he quotes either Psalm 14 or Psalm 53 in the form 'There is no one who is righteous, not even one', where the word 'righteous', so important in Paul's thinking in Romans, does not correspond to the wording of either the Hebrew or the Greek of the two psalms.

Quite often, the influence of scripture is to be detected simply through faint echoes of what had previously been said, as in Paul's celebrated affirmation of a kind of natural theology, expressed in the words 'Ever since the creation of the world [God's] eternal power and divine nature, invisible though they are, have been seen and understood through the things he has made' (Romans 1:20). The words recall in a general way the opening verses of Psalm 19, 'The heavens declare the glory of God, . . .'. Another example is Paul's statement that 'the creation was subjected to futility' (Romans 8:20), which is reminiscent of the beginning of Ecclesiastes, 'vanity of vanities, all is vanity' (1:2). These background influences correspond to something like what I have called the contextual use of scripture.

Sometimes Paul will cite scripture clearly and directly, as in his reference to the commandment not to covet (Romans 7:7), or in the long catena of quotations from the Psalms in Romans 3:10-18. At other times he is content with a free paraphrase, as in the statement of 'what scripture [actually, of course, Moses] says to Pharaoh' (Romans 9:17), which reproduces the sense of both the Hebrew and the Greek of Exodus but not the wording of either. Another technique Paul uses in citation is to combine two separate verses, as he does in fusing together Isaiah 8:14 and 28:16 to form his 'stumbling stone' citation in Romans 9:33. Most striking of all, and of considerable Christological significance, is the way in which Paul is able to take an Old Testament text that clearly refers to the Lord, the one God of Israel, and apply it to Jesus Christ. He does this in Romans 10:13, where he unhesitatingly takes Joel 2:32—'everyone who calls on the name of the Lord shall be saved'—and applies it to the Lordship of Christ. An even more remarkable example of this theological freedom is found in Philippians, where the well-known Christological hymn ends by asserting that 'every knee shall bow' to, and 'every tongue confess', the exalted Christ (Philippians 2:10-11), unmistakably echoing the words of God in Isaiah (45:23), 'To *me* every knee shall bow, every tongue shall swear'.

Paul is certainly no slave to previously received interpretation, as he makes clear in his famous citation (Romans 4:3) of the verse from Genesis (15:6) concerning the faith of

Abraham, that it 'was reckoned to him as righteousness'. Contemporary Jewish understanding seems to have considered the original meaning of 'righteous' here as referring to meritorious deeds (cf. 1 Maccabees 2:52), but Paul gives the verse an entirely different theological significance. When one goes on to note the use that the Epistle of James makes of the same verse (James 2:23), in fact in a much more works-oriented sense, one begins to see something of the hermeneutic flexibility, and creative theological freedom, that the New Testament writers exercise.

So what may we conclude? Have I argued myself into accepting so plastic a role for scripture that almost anything can be done with it? I do not think so. As in the case of Michael Polanyi's illuminating and persuasive account of scientific practice,[22] there are tacit skills of judgement to be exercised, within the oversight of a truth-seeking community and with universal intent, which, though they cannot be reduced to the routine following of a specifiable protocol, are capable of being recognised and respected by those who seek to 'exist within the orbit where the Bible is interpreted'. The necessary flexibility of hermeneutic strategy means that we shall not all be able to agree on every point, but this will not mean that we cannot recognise each other as being engaged in the common task of the search for the truth about the living God that is made known to us through the testimony of scripture. I shall give the last word to Jürgen Moltmann, who tells us,

> I take Scripture as the stimulus to my own theological thinking, not as an authoritative blueprint and confining boundary. It is 'the matter of Scripture' that is important,

22. See Polanyi, *Personal Knowledge*.

not the scriptural form of the matter, even if it is only through that form that we may arrive at the substance. 'God's word is not bound'. . . . The 'meaning' of a text is not exhausted either in its own time or in its interpretation in a new *kairos*. If it is a text belonging to the history of promise and hope, an enticing surplus of 'more' always remains in the text and in its interpretation.[23]

23. Moltmann, *Experiences in Theology*, pp. xxii and 105–106.

CHAPTER THREE

The Universe in a Trinitarian Perspective:
A Theology of Nature

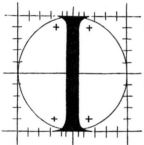 N this chapter I want to travel from physics to metaphysics. I am using 'physics' in its ancient sense of what concerns the nature of things, though concentrating on those aspects of things that are disclosed by the natural sciences, and I am using 'metaphysics' in the sense of a comprehensive, overarching account that embraces the insights of a number of disciplines, each of which views reality from its own particular point of view. The traveller on such an intellectual voyage has a choice of routes and end-points, for metaphysical views are selected and defended for metaphysical reasons and no metaphysical view can claim the degree of coerciveness that would correspond to logical necessity. The relation between physics and metaphysics is a subtle one and there is no inescapable entailment linking the two. Nevertheless, physics constrains metaphysics, rather as the foundations of a building constrain, but do not determine, the edifice that can be erected upon them. The

connection between the scientific concepts of physics and the philosophical or theological concepts of metaphysics, is that of an alogical association, based upon a perceived consonance. I believe that the discoveries of modern science have opened up significant possibilities for fruitful metaphysical construction, which I shall seek briefly to explore. At the same time I shall make what some of my scientific colleagues might think was an over-audacious claim, that a deeply intellectually satisfying candidate for the title of a true 'Theory of Everything' is in fact provided by Trinitarian theology.

The exercise on which I shall engage is somewhat similar to that which in an earlier age might have been called the search for 'vestiges of the Trinity'. Of course, I am not claiming that the world is full of entities stamped 'Made by the Holy Trinity'. God's work of creation is rather more subtle than that. What I shall claim is not that we can infer the Trinity from nature, but that there are aspects of our scientific understanding of the universe that become more deeply intelligible to us if they are viewed in a Trinitarian perspective. It seems to me that it would be perplexing for the Christian believer if no such indications could be found, but I also acknowledge that they will not prove to be of so unambiguous a kind as to force the minds of everyone necessarily into seeing things the way that I do. It is to be expected that God is neither totally hidden nor totally manifested in the works of creation.

The task in hand can fittingly be characterised as the exploration of a theology of nature, in contrast to the more ambitious task of the construction of a natural theology. The latter project aims to infer the existence of a divine Creator as the best explanation of the order and fertility of the universe. I believe that natural theology is a feasible undertaking but, even at its most successful, it can lead to no more than the limited theological insight of the deity as the ground of rational being and fruitful process. My concern in the programme of this book is with a greater theological richness than such a deistic concept can provide. Consequently my purpose in this chapter is to engage in the exercise of using Trinitarian theology to provide an extended context within which to accommodate certain striking features of our current understanding of the cosmos.

There are seven scientifically disclosed features of our universe that I want to consider, the first two of which could also form part of an argument of natural theology.

(1) *A deeply intelligible universe.* It is scarcely surprising that we can understand the world in the everyday way that is obviously necessary for our survival within it. If we could not figure out that it is a bad idea to step off the top of a high cliff, we would not be around for very long. But it does not follow from this that someone like Isaac Newton could come along and, in a great creative leap of the imagination, see that the same force that makes the cliff dangerous is also the force that holds the Moon in its orbit around the Earth and the Earth in its orbit around the Sun, discover the beautiful property of universal inverse-square-law gravity, and so make comprehensible the behaviour of the whole solar system. You may recall that when Sherlock Holmes and Dr Watson first meet, the great investigator shocks the good doctor by feigning not to know if the Earth goes round the Sun or the Sun goes round the Earth. He defends this apparent ignorance by asking what does it matter for his daily work as a detective? Of course, it does not matter at all, but human beings know a great many things that are of no immediate practical relevance.

The development of modern science has shown us that our human ability to understand the universe far exceeds anything that could reasonably be considered as simply an evolutionary necessity, or as a happy spin-off from that necessity. The universe has proved to be astonishingly rationally transparent, and the human mind remarkably apt to the comprehension of its structure. We can penetrate the secrets of the subatomic realm of quarks and gluons, and we can make maps of cosmic curved spacetime, both regimes that have no direct practical impact upon us, and both exhibiting properties that are counterintuitive in relation to our ordinary habits of thought. Our understanding of the workings of the world greatly exceed anything that could simply be required for human survival. The assertion by a so-called evolutionary epistemology of the necessary validity of knowledge acquired in order to support such survival is only a partial insight, insufficient in itself to explain the success of science.

The mystery of the universe's intelligibility is even deeper than that, however, for it has also turned out that it is mathematics that is the key for unlocking these scientific secrets. In fundamental physics it is an actual technique of discovery to look for equations that have about them the unmistakeable character of mathematical beauty. Time and again we have found that it is only equations possessing economy and elegance of this kind that will prove to be the basis for theories whose long-term fruitfulnesss convinces us that they are indeed verisimilitudinous descriptions of physical reality. The greatest physicist whom I have known personally, Paul Dirac, one of the founding figures of quantum theory, once said that it was more important to have mathematical beauty in one's equations than to have them fit experiment! Of course,

Dirac did not mean that empirical success was an irrelevance in physics—no scientist could believe that. Yet, if at first sight one's equations did not appear to fit experiment, there were some possible ways out of the difficulty—maybe you had made a bad approximation and not solved them correctly, or maybe the experiments themselves were wrong. But if the equations were ugly . . . well, there really was no hope for them. Dirac made his many great discoveries, including the existence of antimatter, by a lifelong and highly successful quest for mathematical beauty.

If we stop to think about what is happening when we use abstract mathematics in this way as a guide to physical discovery, we shall see that something very odd indeed is going on. After all, mathematics is pure thought and what could it be that links that thought to the structure of the physical world around us? Dirac's brother-in-law, Eugene Wigner, who also won a Nobel Prize for Physics, once called this 'the unreasonable effectiveness of mathematics'. He said it was a gift that we neither deserved nor understood.

Well, I do not know whether we deserve it, but I would certainly like to understand this remarkable power of mathematics. It would be too intellectually feeble just to say, 'That's the way it is—and a bit of good luck for you people who are good at maths!' If I am to gain that understanding, I shall have to look outside of science itself, for physicists are just glad that this is so and then get on with the task of exploiting the opportunities that it presents. If one is imbued with that thirst for understanding so characteristic of the scientist, one will want to face the fact that science is privileged to explore a universe that is both rationally transparent and rationally beautiful in its deep and accessible order. It does not seem sufficient just to treat this as a happy accident. Scientists frequently speak of the experience of wonder as the reward for all the weary labour involved in their research. Something profound is going on in science's exploration of our deeply intelligible universe that calls for metascientific illumination.

It seems to me that purely naturalistic thinking is unable to cast light on this deep intelligibility, for ultimately it has to treat it as a fortunate but fortuitous fact. Here is the first point in my discussion where a religious view has something extra to offer, for a doctrine of creation suggests that the reason within our minds, and the rational structure of the physical world around us, have a common origin in the rationality of the God whose will is the ground of the existence of both our mental life and our physical experience. Arguments of natural theology often appeal to this fact of cosmic intelligibility. In its turn, a theology of nature can assert that rational transparency and beauty are just what one would expect of a universe that is a divine creation.

In more specifically Trinitarian terms, our scientific ability to explore the rational beauty of the universe is seen to be part of the Father's gift of the *imago Dei* to humankind, and the beautiful rational order of the universe is the imprint of the divine Logos, 'without whom was not anything made that was made' (John 1:3). Whether acknowledged or not, it is the Holy Spirit, the Spirit of truth (John 15:26), who is at work in the truth-seeking community of scientists. That community's repeated experiences of wonder at the disclosed order of the universe are, in fact, tacit acts of the worship of its Creator.

(2) *A universe with a fruitful history.* The universe as we know it originated in the fiery singularity of the Big Bang, some fourteen billion years ago. Initially all was very simple.

At the start the cosmos was just an almost uniform expanding ball of energy. Cosmologists speak with a certain justified confidence about the early universe just because it is so simple a physical system to think about. After fourteen billion years of evolving history, that same universe has become richly diverse and structured, with ourselves the most complex consequences of which we are aware. In one of his books, Holmes Rolston tells us that when an astronomer peers through a telescope at some distant galaxy, he or she should remember that the most complex physical structure we have ever encountered is just six inches this side of the eyepiece—the human brain within the skull of the astronomer.[1] It is a striking fact that that initial ball of energy has become the home of saints and mathematicians. This recognition in itself might raise the question of whether there has not been more going on in cosmic history than science alone can fully express. Might there not be some purpose behind it all?

Of course, the universe's history has been an evolving history, as much on the cosmic scale as it has been in relation to the development of biological life on Earth. As we think about that fact, we may indeed follow the distinguished French biochemist and atheist Jacques Monod in seeing evolutionary process as involving an interplay between chance and necessity,[2] but we need not go on to agree with him in annexing the tendentious adjective 'blind' to the chance half of the process. By 'chance' is not meant the operations of the capricious goddess Fortuna, but simply historical contingency, that this happens rather than that. This particular genetic mutation turns

1. H. Rolston, *Science and Religion*, Temple University Press, 1987, p. 66.
2. J. Monod, *Chance and Necessity*, Collins, 1972.

the stream of life in this particular direction. Had a different mutation occurred instead, a different possibility would have been realised. Not everything that could happen has happened; history necessarily represents only a small selection from the range of cosmic possibility. Chance, therefore, is just a shuffling mechanism for exploring potentiality.[3] Of course, genetic mutations are blind to the pressures of changing environmental circumstances, but natural selection is a subtle means by which life can adapt itself to these circumstances. Theologically this has been understood in a positive sense from the earliest days of evolutionary understanding. It is historically ignorant to suggest that the publication in 1859 of Darwin's *Origin of Species* was greeted either by a unanimous chorus of rejection by the clergy or by a unanimous chorus of acceptance within the scientific community. There were a variety of responses on both sides.[4] Two Anglican clergymen, Charles Kingsley and Frederick Temple, both gave an early welcome to Darwinian insight, seeing that evolution could be theologically understood as the way in which creatures have been allowed by their Creator 'to make themselves'. The God of love has not brought into being a world that is simply a divine puppet theatre, but rather the Creator has given creatures some due degree of creaturely independence.[5] Trinitarian theology does not need to see the history of the world as the performance of a fixed score, written by God from all eternity, but may properly understand it as the unfolding of a grand

3. This has been emphasised especially by A. R. Peacocke, *Creation and the World of Science*, Oxford University Press, 1979, ch. 3.

4. J. H. Brooke, *Science and Religion*, Cambridge University Press, 1991, ch. 8.

5. J. C. Polkinghorne (ed.), *The Work of Love*, SPCK/Eerdmans, 2001.

improvisation in which the Creator and creatures both participate.

Evolution happens within the given necessity of natural law, a point too little attended to by Monod. In recent years the collection of scientific insights called the Anthropic Principle[6] has led us to the surprising conclusion that a universe capable of evolving the complexity of life as we know it is a very special world indeed. While the contingency of evolutionary process is certainly part of the cosmic story, it is only one aspect, and the proper understanding of that story requires the recognition of the 'fine-tuning' of the lawful necessity of the world, which has also been an indispensable element in the fertility of what has been going on. While life did not appear until the universe was about ten billion years old, and self-conscious life when it was almost fourteen billion years old, there is a real sense in which the universe was pregnant with the possibility of carbon-based life almost from the moment of the Big Bang onwards. Its physical fabric was then of the exact form necessary to allow the eventual emergence of life.

This is so remarkable a conclusion that I would like to give two examples of the kind of thinking that lies behind it. Advanced life could evolve only on a planet whose sun was a reasonably steady source of energy over the long period needed to evolve beings with something like the complexity of humans, a process that intrinsically requires three to four billion years to come to full fruition. We know what makes stars burn in this long-lived and reliable way and it turns out to depend upon the strengths of the fundamental forces of nature,

---

6. J. D. Barrow and F. J. Tipler, *The Anthropic Cosmological Principle*, Oxford University Press, 1986; J. Leslie, *Universes*, Routledge, 1989; N. A. Manson (ed.), *God and Design*, Routledge, 2003.

particularly gravity. If these details were significantly different from the way they are in our universe, the stars would be put out of kilter, either burning too feebly to support life, or burning so intensely that they would have burnt themselves out in a mere few million years, far too short a period to be of any use for fuelling the development of life.

Stars have a second indispensable anthropic role to play, for it is only in their interior nuclear furnaces that the chemical raw materials necessary for life—elements such as carbon, oxygen and some twenty others—can actually be made. We are all creatures of stardust, formed from the ashes of dead stars. One of the scientists who unravelled the delicate and beautiful chain of stellar nuclear reactions by which these elements were made was Fred Hoyle. When he saw how this was just possible in a most exquisite way because the basic nuclear forces are just what they are and no different, he is reported to have said, 'The universe is a put-up job'. In other words, it seemed to Hoyle that a process so remarkable could not be just a happy accident. There had to be some Intelligence behind it.

Many scientists were upset when this remarkable specificity of our universe was recognised. They did not like the thought that there was anything special about our world, for they would rather have considered it as just a typical specimen of what a universe might be like. The scientific instinct prefers the general to the specific and it is unnecessarily wary of the particular. A modest relaxation of scientific anxiety in respect of detailed specificity can be achieved by making a reasonable conjecture about physical processes taking place in the very early universe. If one believes, as most physicists do, that the fundamental physical character of the universe is described by a currently unknown Grand Unified Theory

(GUT), then a process called spontaneous symmetry breaking will have taken place a fraction of a second after the Big Bang, reducing that primeval urforce to the effective forces that are seen at work in nature today. This reduction need not have been a literally universal event, for it may have taken different forms in different places. (The process depends upon an instability. It is a bit like a system of pencils all delicately balanced on their points. When slight disturbances break the symmetry, the pencils need not all fall in the same direction.) This means that the universe may be composed of large cosmic domains in which the basic forces operating have different relative strengths. We are unaware of this because another primeval process, called inflation, is believed to have blown up each of these domains to colossal size and our neighbours in this respect are well over the horizon of visibility. We, of course, must be living in that particular domain where the effective forces have just the ratio of strengths that permits the possibility of carbon-based life. This suggestion, which I accept as an entirely credible possibility, only relaxes anthropic specificity to a limited degree.[7] One still requires the primeval GUT to be capable of generating the right sort of inflation (needed to smooth out the universe and prevent its being too turbulent, and also to produce an anthropically necessary balance between expansive and contractive effects), and also the right sort of symmetry breaking, that is to say it must be of a kind that is capable of reducing in some domain to the anthropic fine-tunings that are actually observed. Moreover, the laws of nature must be quantum-mechanical and include gen-

---

7. J. C. Polkinghorne, *Faith, Science and Understanding*, SPCK/Yale University Press, 2000, § 5.2.

eral relativity, conditions that are anthropically necessary but not logically required. The universe would still be far from being just 'any old world'.

Those who wish to try to defuse anthropic uniqueness more drastically are driven to the prodigal supposition that there is a truly vast portfolio of distinctly other universes, all with greatly different sorts of natural laws and circumstances and all, of course, inaccessible to us. Our universe is then just the one where fortuitously carbon-based life is possible—a winning ticket in a multi-cosmic lottery, one might say. This proposal of a prodigious multiverse is not a scientific suggestion but a metaphysical speculation, a way to accommodate anthropic fine-tuning within a recklessly enlarged naturalism. It seems to me that a much more economic understanding is offered by the belief that there is only one universe, which is the way it is because it is indeed not 'any old world' but a creation that has been endowed by its Creator with just those finely tuned laws that have enabled it to have a fruitful history.

Like all metaphysical discussion, the argument is not of a logically coercive, knock-down kind, but for me it is coherent and intellectually satisfying. It could properly form part of a cumulative case for natural theology and it certainly fits with admirable consonance into a theology of nature. It is important to recognise that belief in God the Creator does a great number of other pieces of explanatory work beyond anthropic issues, such as enabling one to understand the universe's deep intelligibility as well as indicating the origin of the widely attested human experience of encounter with the sacred. On the other hand, the many-universes hypothesis seems to do only one piece of explanatory work, namely granting relief from taking seriously the possibility of theism. I do not think that

William of Occam, whose logical 'razor' was designed to cut away speculative excess, would have been happy with such a rash multiplication of entities.

Finally, in relation to the evolving character of fruitful cosmic history, we need to note that it also turns out that evolutionary understanding affords a way in which scientific insight can offer religious belief some modest help with what is surely the latter's greatest perplexity. I refer, of course, to the presence of evil and suffering in the world, a problem that presents a theology of nature with its most serious challenge. A creation allowed to make itself can be held to be a great good, but it has a necessary cost not only in the blind alleys and extinctions that are the inescapable dark side of the evolutionary process, but also in the very character of the processes of a world in which evolution can take place. The engine driving biological evolution is genetic mutation and it is inevitable in a universe that is reliable and not capriciously magical, that the same basic biochemical processes which enable germ cells to produce new forms of life will also allow somatic cells to mutate and become malignant. That there is cancer in creation is not something that a more competent or compassionate Creator could easily have eliminated, but is the necessary cost of a creation allowed to make itself. The more we understand scientifically the process of the world, the more it seems closely integrated—a package deal from which it is not possible in a consistent way to retain the 'good' and remove the 'bad'. I do not for a moment believe that this insight eliminates all the anguish and perplexity that we feel at the evil and suffering in the world, but it does suggest that its presence is not gratuitous.

The depth of the problem posed by the demands of theodicy is only adequately met in Christian thinking by a Trinitarian understanding of the cross of Christ, seen as the event in which the incarnate God truly shares to the uttermost in the travail of creation. Jurgen Moltmann[8] has led us to understand that the Christian God is not just the compassionate spectator of the suffering of creatures, but the Crucified God, who is truly the 'fellow sufferer who understands' (Whitehead), the One who is creation's partner in its pain.

(3) *A relational universe.* Newtonian physics pictured the collisions of individual atoms as taking place within the container of absolute space and in the course of the unfolding of a universal absolute time. Einstein's theory of special relativity showed that judgements of simultaneity and of the elapse of time are not absolute, but are relative to the state of motion of the observer. In what many would regard as his greatest discovery, Einstein then went on to develop the theory of general relativity, showing that space, time and matter are closely interconnected in a kind of integrated package, in which matter curves spacetime and spacetime curves the paths of matter. The cosmic 'container' and its contents are not separable, but intimately linked with each other.

Later still, Einstein, through the discovery of what is called the EPR effect,[9] showed that quantum theory implied that once two quantum entities have interacted with each other, they remain mutually entangled however far they may eventually separate. This counterintuitive togetherness-in-separation (non-locality) seemed so 'spooky' to Einstein that he supposed it indicated that there was something incom-

8. J. Moltmann, *The Crucified God*, SCM Press, 1974; see also P. Fiddes, *The Creative Suffering of God*, Oxford University Press, 1988.

9. See J. C. Polkinghorne, *Quantum Theory: A Very Short Introduction*, Oxford University Press, 2002, ch. 5.

plete in the quantum account. However, the beautiful experiments of Alain Aspect and his collaborators have shown us that non-locality is indeed a property of nature. It has turned out that even the subatomic world cannot be treated atomistically! Twentieth-century science has revealed a deep-seated interconnectivity present in the fabric of the physical world.

I do not believe that metaphysical thinking has yet fully absorbed these developments or come to terms with the fact that localised individuality is no longer an unproblematic concept. Here is an opportunity for important further developments in ontological thinking, which we may hope will be grasped in the course of the twenty-first century. It is striking that so methodologically reductionist a subject as physics has pointed us in this relational and holistic direction. This tendency is surely reinforced by chaos theory's discovery that at the macroscopic level of physical process there are many systems that are of such exquisite sensitivity to the details of their circumstance that they cannot properly be isolated from the effects of their environment. The slightest disturbance will totally change their future behaviour. Of course, actual experimental exploration has concentrated on the investigation of those situations where isolatability is an acceptable idealisation. Otherwise the scientific task would have been impossible, for one would have had to understand everything before one could understand anything. However, the general character of physical reality seems to correspond to a weblike character of interconnected integrity. An ontological possibility that needs serious consideration is that the detailed character of the laws of nature that we have formulated on the basis of isolatable experimentation is no more than what one might call a 'downward-emergent' approximation to some more holistic account of physical reality, so that what we presently believe we know is only really valid in the special circumstances that an effective degree of separation is a good approximation to the situation.[10] Section 6 will suggest a possible form that such an enlarged account might take.

With physics moving in a more holistic direction, we might expect by analogy to need to challenge the individualistic atomism that is so characteristic of contemporary thinking about human nature. If electrons are counterintuitively entangled with each other, we may need to contemplate the possibility that persons participate in some greater solidarity than atomised Western society is able to recognise.[11] Such an insight is surely consonant with the Christian understanding of the community of the faithful as the Body of Christ, constituting a web of relationality vastly more comprehensive than the one-to-one exchange of I and Thou.

These remarkable developments in relational and holistic thinking that are taking place within the fold of science are deeply congenial to Trinitarian ways of thought. They by no means 'prove' the Trinity, but they are profoundly consonant with a theology of nature that sees the relation of perichoretic exchange between the divine Persons as lying at the heart of the Source of all created reality. One could paraphrase the title of John Zizioulas's insightful book on Trinitarian theology, *Being as Communion*,[12] by the phrase 'Reality is relational.'

10. J. C. Polkinghorne, *Science and Christian Belief/The Faith of a Physicist*, SPCK/Princeton University Press, 1994, pp. 25-26.

11. The insights of some forms of depth psychology, and of tribal societies, may be helpful here.

12. J. Zizioulas, *Being as Communion*, Darton, Longman & Todd, 1985.

(4) *A universe of veiled reality*. Although the world appears so clear and reliable in terms of our everyday experience of it, quantum theory has shown us that it is cloudy and fitful at its subatomic roots. There are still many unresolved disputes about what is the most satisfactory meta-interpretation of quantum physics. A number of alternative proposals, none altogether free from difficulty, are on offer,[13] but whatever interpretation of quantum theory one may choose to embrace, common to all is the realisation that Heisenberg's uncertainty principle sets limits to our epistemic access to what is going on. Whether this is due to an unfortunately necessary ignorance of hidden detail, as David Bohm believed, or whether it is an intrinsic ontological property of subatomic reality, as conventional quantum theory of the kind stemming from the Copenhagen interpretation of Niels Bohr asserts, is still an unresolved issue, though the great majority of physicists side with Bohr against Bohm. The empirical equivalence of the predictions of Bohmian theory and of conventional quantum theory imply that the question is metaphysical in character, lying beyond the simple possibility of experimental settlement. If one adopts the view, first commended by Werner Heisenberg, that one should consider the quantum world as a realm of potentialities which can only become fleeting actualities as a result of experimental interrogation, then its reality must take an idiosyncratically veiled form. These issues have been particularly well elaborated by the French philosopher-physicist Bernard d'Espagnat,[14] but much work on the nature of quantum ontology still remains to be done. One may make

13. Polkinghorne, *Quantum Theory*, chs. 3 and 6.
14. B. d'Espagnat, *Reality and the Physicist*, Cambridge University Press, 1989.

two further comments that seem to be of particular relevance to theology.

Most physicists remain convinced that 'reality' is the correct term to apply to quantum entities. The concept of electrons is not simply a convenient manner of speaking; they are really there. The necessity of the predicate 'veiled' to qualify that reality reminds us that it must be encountered on its own terms and in accordance with its Heisenbergian uncertainty. In other words, there is no universal epistemology. We can only know the quantum world in terms respectful of its veiling, and it would be epistemically disastrous to try to insist on the Newtonian clarity that we can often attain in the macroscopic world of everyday phenomena. If that epistemic specificity is true of subatomic physics, it is surely even more important to recognise a similar truth in relation to the knowledge of God. This is a point that Thomas Torrance has been particularly helpful in emphasising: 'How God can be known must be determined from first to last by the way in which He is actually known'.[15] Theology cannot operate under the yoke of a basically agnostic epistemology imposed upon it a priori. It is the Christian testimony that God is most fully to be known in meeting with the One God whose triune reality is Father, Son and Holy Spirit. Of course this way of thinking is counterintuitive, just as so much of quantum theory is counterintuitive, but, just as in the case of quantum mechanics, that novel pattern of thought is forced upon us by the reality encountered and it does not arise from fanciful or unconstrained speculation.[16]

15. T. F. Torrance, *Theological Science*, Oxford University Press, 1969, p. 9.
16. Polkinghorne, *Science and Christian Belief/The Faith of a Physicist*, pp. 154–56.

The other comment relates to the ground on which the reality of quantum entities is to be defended, even including the reality of confined entities such as quarks and gluons that we believe will never be observable individually. The basis for belief in these unseen realities lies in the way in which that belief makes sense of great swathes of more accessible physical phenomena (technically, the structure of the hadronic spectrum and the results of deep inelastic scattering). In the mind of the physicist it is the resulting deep intelligibility that is the basis for ontological belief. Once again, the analogy with theology is clear enough. Belief in the unseen reality of God can properly be defended on the basis of the insightful understanding that it yields in relation to great swathes of spiritual experience, particularly in relation to the gospel record and its testimony to Jesus Christ, and in relation to the continuing worshipful and sacramental experience of the Church.

(5) *A universe of open process.* It sometimes seems that people outside the scientific community still suppose that the universe that science describes is no more than a gigantic piece of cosmic clockwork. In actual fact, the twentieth century saw the death of a merely mechanical view of the world. Its demise came about through the discovery of widespread intrinsic unpredictabilities present in physical process, first at the subatomic level of quantum theory, and then at the everyday level of those exquisitely sensitive systems that have been given the ill-chosen name of 'chaotic'. (The adjective was ill chosen because chaos theory actually involves a subtle interplay between order and disorder, future behaviour being unpredictable but not totally haphazard.) It is important to emphasise the intrinsic character of these unpredictabilities. They do not arise from deficiencies in experimental technique, or from a lack of

calculating power, but are limits in principle that cannot be overcome.

All scientists would agree that these are highly significant and surprising discoveries, but the matter becomes more contentious when we go on to discuss what they might actually imply for the process of the world. Unpredictability is an epistemological property and there is no inevitable connection between epistemology and ontology. What connection we make is a matter of metaphysical choice and philosophical contention. In particular, questions of the nature of causality are always ultimately metaphysical in character, as the unresolved dispute between Bohm and Bohr about whether quantum theory should be considered deterministic or indeterministic makes only too clear. Different people will adopt different strategies.[17] As a scientist, my instinct is to adopt a realist stance, that is to believe that what we know is a reliable guide to what is the case. We should trust well-winnowed knowledge not to mislead us. I have encapsulated this strategy in a slogan that I coined, and of which I am rather fond: 'Epistemology models Ontology', what we know is a reliable guide to what is the case. After all, why go to all the trouble involved in doing science if one does not believe that thereby we are learning what the physical world is actually like?

If you take this realist view, then unpredictabilities will not be seen as unfortunate epistemological deficits but rather as signs of an actual ontological openness to the future. The

17. For surveys, see J. C. Polkinghorne, *Belief in God in an Age of Science*, Yale University Press, 1998, ch. 3; *Faith, Science and Understanding*, chs. 6 and 7; R. J. Russell, N. Murphy and A. R. Peacocke (eds.), *Chaos and Complexity*, Vatican Observatory, 1995; R. J. Russell, P. Clayton, K. Wegter-McNelly and J. C. Polkinghorne (eds.), *Quantum Mechanics*, Vatican Observatory, 2001.

vast majority of quantum physicists take this view when they side with Bohr against Bohm and see the uncertainty principle as a principle of indeterminacy and not merely of ignorance. I have proposed that we should do the same with chaos theory, regarding the deterministic Newtonian equations from which the discussion began as no more than emergent downward approximations to a more subtle and more supple physical reality. By that claim of openness, of course I do not mean that the future becomes some random lottery, but that the causes that bring it about are more than simply the exchanges of energy between constituents that a conventional science describes. What then might be the nature of such additional causal principles? First, the unisolatable character of chaotic systems implies that this new form of causality will be holistic, referring to the behaviour of the system as a whole and not simply to its separate parts. In a word, it will be top-down rather than bottom-up. I also suggest that these principles will be concerned not with energy but with what one might call the input of information, that is the specific generation of patterns of behaviour. The unpredictable future possibilities of a chaotic system differ from each other in precisely this way; they correspond essentially to the same energy but to different patterns in which the energy flows. This emphasis on pattern and information is an issue to which I shall return. Before I do so, let us consider what the picture of a universe of open process might mean for a Trinitarian theology of nature. Two insights seem particularly relevant.

First, the picture implies that creation is a world of true becoming and not a world of static being. Returning to a previous musical metaphor, cosmic history is an unfolding improvisation and not the performance of an already written score. Since God knows things as they truly are, this will surely imply that God knows such a world in its becomingness. In other words, we are encouraged to embrace the notion of a divine dipolarity with respect to time, holding the eternal and the temporal in mutual complementarity, of the kind that we have already discussed in relation to the understanding of the nature of scriptural prophecy (pp. 53-55). I shall return to this point again in Chapter 4.

The second point arises from the fact that a scientific way of characterising the fruitfulness of open physical process is to say that novelty emerges in regimes that are 'at the edge of chaos'. Here order and openness so interlace that the state of affairs is neither so rigid that that nothing really new can ever come about, nor so haphazard that nothing new can ever persist. A Trinitarian theology of nature has some resonance with this insight. The Father is the fundamental ground of creation's being, while the Word is the source of creation's deep order and the Spirit is ceaselessly at work within the contingencies of open history. The fertile interplay of order and openness, operating at the edge of chaos, can be seen to reflect the activities of Word and Spirit, the two divine Persons that Irenaeus called 'the hands of God'.[18] There is a Trinitarian rhythm of sustaining-redeeming-sanctifying that sees creaturely ennoblement as issuing from rescue, a concept expressed in the paradoxical cry *O felix culpa!* The insight of creativity at the edge of chaos could be seen as a pale reflection of this pattern.

18. E. Osborn, *Irenaeus of Lyons*, Cambridge University Press, 2001, pp. 89-93.

The insight of a theology of nature discussed in this section differs in its character from that discussed in section 3. The latter suggested a connection between the inner nature of God (the immanent Trinity, as theologians say) and the relational character of creation. Here an analogy is being drawn between the external activities of God (the economic Trinity) and the scientifically observed processes of creation.

(6) *An information-generating universe.* I believe that we are on the threshold of very interesting new developments in basic scientific understanding. Through computer simulation and some other techniques, we are just beginning to learn something about the detailed behaviour of genuinely complex systems. It turns out that they display quite astonishing propensities for the spontaneous generation of patterns of large-scale order. A very striking example of this has been given by Stuart Kauffman in his computerised investigations into the behaviour of what are called Boolean nets of connectivity 2.[19] The details are not important for our present purpose, but the results are extremely intriguing. One can picture the essence of what is going on by thinking of a large array of electric light bulbs, each of which is either on or off. Every bulb is correlated with two other bulbs somewhere else in the array. The system develops in steps and the form of the correlation implies that the state of a bulb at the next step depends upon the present states of its correlates. If the net contains 10,000 elements, there are about 10^{3000} states of illumination in which the array might be found. However, it turns out that a net started off in a random configuration does not just twinkle away haphaz-

19. S. Kauffman, *At Home in the Universe*, Oxford University Press, 1995, ch. 4.

ardly for ever, but very soon settles down to cycling through only about a hundred different patterns of on/off illumination. This represents the spontaneous generation of an altogether astonishing degree of order.

At present these matters are not well understood, the subject being at that natural-history stage of investigation in which very striking and unexpected behaviours are observed but their origin remains fundamentally mysterious. I hope that this state of ignorance will not continue for long and I expect that the science of the twenty-first century will be characterised by its making dynamic pattern, and the information that specifies that pattern, a fundamental category in scientific vocabulary, alongside the traditional concepts of matter and energy. We might then expect to be able to combine this new emphasis on patterned behaviour with the causal openness proposed in the previous section, thereby adding to the portfolio of our causal imagination the concept of *active information*, a dynamical pattern-forming propensity that operates in a holistic way on totalities rather than separably on constituents. Here we may see a *glimmer*—I say no more than that— of how it might be that we enact our chosen patterns of behaviour as intentional agents. Even at this conjectural stage of discussion, the proposal obviously requires more consideration than I can give it now, but it is something that I have tried to lay out in somewhat more detail elsewhere.[20] If this way of thinking is correct, it will be significant not only in relation to human agency, but also in relation to how theologically we may conceive of the action of divine providence in creation.

20. See Polkinghorne, *Belief in God in an Age of Science*, ch. 3; *Faith, Science and Understanding*, chs. 6 and 7.

God may be seen as interacting with creation by the input of information within its open history.

Certainly, one consequence of the picture I have been developing is that science's description of physical process is not drawn so tight as to condemn God to the non-interactive role of a deistic spectator. Another consequence is that, if the locus of agential action is always within the cloudiness of unpredictability, it follows that, though that action is real, it will always be hidden to a necessary degree. What is going on cannot be analysed exhaustively and itemised into components, so that one might assert that nature did this, human will did that, and divine providence did the third thing. Providence may be discernible by the eye of faith, but it will not be exhibitable by experiment. God is indeed a *deus absconditus*, a hidden deity.

This last insight seems to me fully compatible with the account that Christian theology has sought to give of the working of the Spirit, often discretely and hiddenly operating on the inside of creation, guiding and influencing its history but not inevitably manifested in some overwhelming and unambiguous way.[21] God acts within the open grain of nature and not against it. God interacts with creatures but does not overrule them, for they are allowed to be themselves and to make themselves. It follows from this that not everything that happens will be in accordance with God's direct will. The divine sharing of the causality of the world with creatures will permit the act of a murderer or the incidence of a cancer, though both events run counter to God's good desires. Involved in creation

21. J. C. Polkinghorne and M. Welker, *Faith in the Living God*, SPCK/Fortress Press, 2001, chs. 5 and 6; J. V. Taylor, *The Go-Between God*, SCM Press, 1972.

is a divine kenotic act of self-limitation that truly allows creatures to be themselves and to make themselves.[22] How the balance is struck between what God does and what creatures do is the age-old theological problem of grace and free will, now written cosmically large.

(7) *A universe of eventual futility.* On the largest possible scale, the history of the universe is a continuing contest between two opposing principles: the explosive force of the initial Big Bang, driving matter apart and augmented, perhaps, by the effects of what has come to be called 'dark energy', and the contractive force of gravity, pulling matter together. Presently these effects are very evenly matched and, while cosmologists currently favour the possibility that expansion will predominate, it would be prudent for thinking about the significance of the long-term cosmic future to take into account both possibilities. If expansion prevails, the galaxies will continue to fly apart for ever, slowly cooling and decaying until the world ends in a long-drawn-out dying whimper. If, on the other hand, contraction prevails, the present expansion will one day be halted and reversed and the world will end in a bang, as the universe collapses back into the melting pot of the Big Crunch. Either way, the cosmos is condemned to eventual futility. It is as certain as can be that carbon-based life will everywhere prove to have been a transient episode in its history.

These reliable but bleak prognostications do not support any notion of long-term evolutionary optimism, the idea of a total and lasting fulfilment to be found within the unfolding of present cosmic process alone. For a theology of nature they

22. See the essays in Polkinghorne (ed.), *The Work of Love.*

raise the issue of whether this eventual futility is compatible with the claim that the universe is a creation, the expression of the benevolent will of its Creator.[23]

Personally, I do not think that the knowledge of the universe's death on a timescale of very many tens of billions of years raises any greater theological difficulties than does the even more certain knowledge of our own deaths on timescales of tens of years. If there is hope, either for the universe or for us, it can only lie in the eternal faithfulness of God—a point that Jesus made clearly in his discussion of these matters with the Sadducees (Mark 12:18–27). Of great importance here are the various New Testament passages that speak in an astonishing way of the cosmic significance of Christ (John 1; Romans 8; Colossians 1). Also important, I believe, is the witness of the empty tomb, for the fact that the Lord's glorified body is the transmuted form of his dead body speaks to me that in Christ there is a destiny not for humanity only, but also for matter, and so for creation as a whole.

I shall return to these issues in more detail later (Chapter 6). What is at stake is the fundamental issue of whether the universe is a cosmos or a chaos. Does the universe make total sense, both now and always, or is its history ultimately 'a tale told by an idiot, full of sound and fury, signifying nothing'? The distinguished theoretical physicist and staunch atheist Steven Weinberg, surveying the scene from his naturalistic point of view, concluded, in the light of eventual cosmic futility, that the more he understood the universe, the more it

23. For a much fuller discussion of eschatological issues, see J. C. Polkinghorne and M. Welker (eds.), *The End of the World and the Ends of God*, Trinity Press International, 2000; J. C. Polkinghorne, *The God of Hope and the End of the World*, SPCK/Yale University Press, 2002.

seemed pointless to him.[24] He could only face it with a kind of heroic defiance. There is a certain nobility in that bleak point of view, but I do not believe that we are driven to embrace it. Yet if we are to be able with intellectual integrity to hold to a more hopeful view, I think this requires the kind of developed theistic system of belief that Trinitarian theology provides. This is the kind of overarching understanding that is necessary if we are to be able to recognise that our world is indeed a cosmos after all, and that is why I believe that we should look at the universe from a Trinitarian perspective.

24. S. Weinberg, *The First Three Minutes*, André Deutsch, 1977, p. 149.

CHAPTER FOUR

Theological Thickness

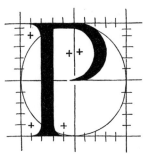REVIOUS chapters have indicated how the Book of Nature and the Book of Scripture can be read in ways that are both supportive of Christian belief and also illuminated by that belief. The level of discourse has been similar to that which in theoretical physics would be called 'phenomenological'. That is to say, significant aspects have been identified and proposals have been made for the broad categories within which understanding of the phenomena can most satisfactorily be located. When scientists are investigating a new physical regime, a phenomenological exploration of this kind is often the necessary preliminary. It offers a sketch-map of the territory, identifying some of its most notable features but falling short of giving a detailed understanding of the intellectual topography and ecology of the domain. Albert Einstein's explanation of the photoelectric effect (1905), and Niels Bohr's model of the atom (1913), provided this kind of fruitful initial insight into

aspects of what was going on in the realm of quantum phenomena.[1] However, the ideas that these bold pioneers proposed were very ad hoc in their character, and how they should be integrated with the previously highly successful concepts of Newtonian and Maxwellian physics was obscure and uncertain. The theorists had to press on to seek a fully articulated and integrated account of quantum mechanics. When the moment of enlightenment came, it did so with remarkable rapidity and fullness of achievement. In the *anni mirabiles* of 1925 and 1926, the essential character of modern quantum theory was brought to birth through the initiating discoveries of Werner Heisenberg and Erwin Schrödinger, and in subsequent important elucidations due to Paul Dirac and Max Born.

In a similar fashion, theologians cannot rest content simply with the direct insights of natural theology and biblical theology. They are driven to seek the integration of this material into the more fully articulated scheme of a systematic theology. The theology of nature discussed in the preceding chapter was a step in this direction, but obviously there is much more to be done. The systematic task is both essential and yet beyond the power of human completion. In physics we are seeking to understand a physical world that, for all the signs of its inexhaustible richness, is one that we transcend and can put to the experimental test. At least within well-defined and well-winnowed regimes, we can expect to gain the kind of extensive and stable success that eventually proved possible in the case of quantum theory.[2] There is a striking contrast in

1. See J. C. Polkinghorne, *Quantum Theory: A Very Short Introduction*, Oxford University Press, 2002, ch. 1.

2. Even that achievement can have its limitations. In the case of quantum theory, we know how to do the sums, and the answers prove to be in impressive

this respect between science and theology, for in the latter case our concern is with the infinite reality of the God who transcends us and whose nature will never adequately be encapsulated within the finite limits of any human rational scheme. An apophatic recognition of the ultimate mystery of the divine is an essential component in a faithful theology. Yet, if God has acted to make the divine nature known through the character of creation and by revelation within history, as Christians believe to be the case, then kataphatic utterance is also a necessary part of the theological enterprise. Theologians should neither be too rationally over-confident nor totally tonguetied.

The attempt to articulate this utterance has been an activity of the Church over two millennia. Trinitarian and incarnational thinking arose precisely through the struggle in the early centuries to integrate Jewish, apostolic and ecclesial experience of God into a single account which, if it did not attain the coherence and adequacy that is possible for a mature scientific theory, would at least take the discussion beyond the immediately phenomenological. The result of this struggle, continued down the subsequent centuries, has been a 'thickness' of theological thinking that is worthy of deep respect and attention. This fact constitutes the principal reason why I feel that I must take a developmental attitude to the relationship between science and theology (pp. 26–29). When I wrote my Gifford Lectures I explained that I did so to explore 'whether the strange and exciting claims of Christianity are tenable in a scientific age'. I also commented that 'A scientist

---

agreement with experiment, but there still remain interpretative questions that are matters of unresolved dispute.

expects a fundamental theory to be tough, surprising and exciting'.[3] You could not get a more fundamental theory than that which Trinitarian theology offers to us.

This chapter will further explore the promise of the Trinitarian perspective in relation to our understanding of the nature of God. It is a task that I, as a theoretical physicist with a serious concern for theology, but without a lifetime of study devoted to that subject, undertake with some diffidence and trepidation. Much writing on the Trinity is formidably technical in its character, and is often astonishingly self-confident in its assertions.[4] As people speak of begetting and procession, of filiation and spiration, of perichoresis and appropriation, sometimes one may perceive something of the motivation that lies behind these utterances, but at other times one is driven to wonder 'How do they know?' Perhaps a scaled-down analogy to the ambition of theological discourse about the divine infinity can be provided from within the experience of the fundamental physicist by considering the confident way

3. J. C. Polkinghorne, *Science and Christian Belief/The Faith of a Physicist*, SPCK/Princeton University Press, 1994, p. 1. For a dialogue between a scientist and a theologian about these issues see J. C. Polkinghorne and M. Welker, *Faith in the Living God*, SPCK/Fortress Press, 2001.

4. A convenient introduction to Trinitarian thought can be found in C. M. LaCugna, *God For Us*, HarperSanFrancisco, 1991. There has been much recent writing expressing a recovery of the importance of the Trinitarian perspective. Significant contributions include: D. Brown, *The Divine Trinity*, Duckworth, 1985; C. E. Gunton, *The Promise of Trinitarian Theology*, T&T Clark, 1991; Gunton, *The One, the Three and the Many*, Cambridge University Press, 1993; R. W. Jenson, *Systematic Theology*, vol. 1: *The Triune God*, vol. 2: *The Works of God*, Oxford University Press, 1997 and 1999; G. Newlands, *God in a Christian Perspective*, T&T Clark, 1994; R. Williams, *On Christian Theology*, Blackwell, 2000. There are interesting analogies, worthy of further consideration, between theology's concern to preserve both divine unity and the true distinctiveness of the Persons and science's concern to strike a balance between interconnectedness and a degree of experienced separability in its description of the physical world.

in which quantum cosmologists talk about the extremely early universe and about the proliferating cosmic sequences of a hypothesised quantum multiverse.[5] Their talk is both fascinating and precarious. The pretty arabesques that the quantum cosmologists perform are executed on the thinnest of intellectual ice and to the sound of cracking. The problem is that currently the two relevant fundamental scientific theories—quantum theory and general relativity—are imperfectly reconciled with each other, and we are not sure how they are to be combined in a wholly consistent fashion. As a result, quantum cosmologists offer us a wealth of exciting notions (such as 'imaginary time' or 'baby universes'), but also a variety of competing and mutually incompatible proposals.

The theological disputes about the Trinity, such as the millennium-old controversy between the East and the West about the procession of the Holy Spirit (the 'Filioque' clause in the Nicene Creed), have a somewhat similar air of sharp disagreement about matters almost ineffable in their character. While struggling with these matters, I was rather encouraged to read the conclusion of the philosophical theologian Keith Ward—actually in this case writing about puzzles concerning how time and eternity are related to each other—that 'The limits of rational theology verge on phantasy, and, as our normal concepts begin to break and fall away, we find ourselves in a logical vacuum where reason and imagination become confused'.[6] In theology we have to do the best we can. Theologians are certainly not to be denied access both to prosaic and to poetic modes of expression, and speculative thinking will

5. See M. Rees, *Before the Beginning*, Simon and Schuster, 1997.
6. K. Ward, *Rational Theology and the Creativity of God*, Blackwell, 1982, p. 167.

have its role, as indeed it does in science. Yet it is important to be aware of the kind of warrant that can be given for what is being said, and to distinguish between what is rather directly rooted in interpreted experience and what is at some degree of hypothetical remove from such experience.

While such caveats must be borne in mind, the task of engagement with theological thickness is, nevertheless, one that must be attempted. There are five topics that seem particularly relevant to a Trinitarian engagement with science and religion. The first of these is:

(1) *God in relation to creatures.* Christian theology has always strongly resisted a pantheistic identification of God and nature, of the kind that the philosophy of Benedict Spinoza endorsed. In doing so, theology has run counter to an inclination that is to be found in the thinking of a number of scientists who have sought to add a religious gloss to their feeling of wonder at the deep rational order and fruitful history of the universe. Einstein rejected belief in a personal God but he often liked to speak about 'the Old One', employing the term as a cipher for the intellectually satisfying fundamental patterns of the physical universe that induced in him a true feeling of awe. He said that if he had a God, it was indeed the God of Spinoza, a thinker whom he greatly admired and concerning whom he once wrote a poem that begins with the line 'How much do I love that noble man'.[7]

This attitude will not do for Christianity. Its God is not a World Principle, embodied in the cosmos and so both coming into being with the origin of the universe and also fading away into nothingness when that universe eventually draws to its

7. M. Jammer, *Einstein and Religion*, Princeton University Press, 1999, p. 43.

dying close. The Christian God is the Ground of the hope of a destiny beyond death, both for human individuals and for the cosmos itself. This thought alone requires that Christian theology make a sharp distinction between creation and its Creator, whose purposes extend beyond the ultimately futile history of the present world.

This insight is further supported by human worshipful encounter with the sacred. The numinous element in that experience testifies to the presence of the divine Other, the One who stands over against humanity in mercy and in judgement. Yet there is also a complementary element in that experience, witnessed to in the most intense way by the testimony of the mystics of all ages and traditions, which speaks of the closeness of the divine to the human worshipper. Paul, in his 'university sermon' in the Athenian Areopagus, spoke of the God 'in whom we live and move and have our being' (Acts 17:28). Theologians need to be able to speak both of divine transcendence and of divine immanence, experienced in these ways.

It is widely recognised by many today that the theological way of thinking which may be called 'classical theism', expressed in the West by the tradition running from Augustine to Aquinas and on through the thought of people like Calvin, laid too great a stress on divine transcendence. Its picture of a God wholly outside created time, acting on creatures but not at all acted on by them, so distanced the Creator from creation as to seem to imperil the fundamental Christian conviction of the love of God for that creation. A great deal of recent theological thinking has sought to redress the balance between divine transcendence and divine immanence.

Among scientist-theologians, a popular way in which to seek to do so has been supposed to lie in panentheism, the belief that 'the Being of God includes and penetrates the whole universe, so that every part of it exists in Him, but His Being is more than, and not exhausted by, the universe'.[8] The language employed is not free from ambiguity. The word 'penetrates' need imply no more than the immanent divine presence to the created universe, but the word 'includes', placed in parallel with it, seems to point to some closer form of ontological relationship. Similar ambiguity attaches to how precisely we are to construe existing 'in Him' (which, when Paul the Jew used a similar phrase in his Athenian address, surely meant no more than an assertion of divine immanence), together with God's being described as being 'more' than the universe, which seems to imply that the universe is, in fact, part of the divine being. Arthur Peacocke has denied that panentheism treats the world as part of God, but when he writes that 'God is in all the creative processes of his creation and they are all equally "acts of God" for he is at all times present and active in them as their agent',[9] these words seem to amount either to an endorsement of classical theology's assertion of an all-pervading divine primary causality co-present within the nexus of secondary creaturely causalities (an idea whose coherence is far from obvious), or to the incorporation of creation within the divine in some way. More recently, Peacocke has explicitly dissented from process theology's picture of equal divine participation in all events.[10] Of course, it is process theology that underlies Ian Barbour's acceptance of the

8. F. L. Cross and E. A. Livingstone (eds.), *The Oxford Dictionary of the Christian Church*, 3rd edition, Oxford University Press, 1997, p. 1213.

9. A. R. Peacocke, *Creation and the World of Science*, Oxford University Press, 1979, p. 204.

10. A. R. Peacocke, *Theology for a Scientific Age*, enlarged edition, SCM Press, 1993, p. 372.

panentheistic idea, but one must recognise that process thinking results in a highly qualified form of divine association with
creation. Although the divine lure is part of every actual occasion, seeking to draw the event's outcome in a preferred direction, the particular concrescent result that actually occurs
is determined by the occasion itself, implying a considerable
degree of effective detachment of creation from the God who
exercises persuasion but possesses no direct power.

Recently, Philip Clayton has given a careful and extended
defence of panentheism from the point of view of a philosophical theologian.[11] But even his discussion is not free from
an acknowledged degree of semantic plasticity, as the occa-
sional use of phrases like 'in a sense' indicates.[12] From a philosophical point of view, one of the considerations held to point
to panentheism is the need for the divine infinity to be absolutely inclusive. Clayton says that 'it turns out to be impossible
to conceive of God as fully infinite if he is limited by something outside himself'.[13] No doubt that would be true if the
limitation were really externally imposed upon God, but what
if the limitation is one that is freely internally accepted by the
divine Love as the necessary cost of holding in being a creation that has been endowed by its Creator with the freedom
that allows it to be itself and to make itself? A very important contemporary theological insight is the recognition that
the act of creation is an act of divine kenosis, precisely involv-

11. P. D. Clayton, *God and Contemporary Science*, Edinburgh University Press,
1997, esp. ch. 4. For a critique, see J. C. Polkinghorne, *Faith, Science and Understanding*, SPCK/Yale University Press, 2000, § 5.3.

12. Ibid., pp. 90, 94, 99, 100; see also p. 102: 'we are composed, metaphori-
cally speaking at least, out of God'.

13. Ibid., p. 99.

ing a self-limitation of this sort.[14] It involves a kind of divine 'making way' for the existence of the created other. Kenosis is the fulfilment of God's power, not its curtailment, for it is the expression of the Creator's love for creation. This kenotic insight is of particular significance in relation to questions of theodicy, for it implies that not everything that happens—neither the act of a murderer nor the incidence of a cancer—is brought about directly by God or is in accordance with the divine will. Here is a problem to which panentheistic thinkers seem to have paid too little attention. The more closely God is identified with creation, the more acute become the problems posed by the existence of evil within that creation.

The classic Christian doctrine of the distinction between Creator and creation has come under question before in the course of the Church's history. In earlier centuries the threat came from neo-Platonism's alternative view, which depicted the world as being an emanation from the divine, existing at the outermost and attenuated fringe of deity. Panentheism represents a kind of modern version of emanationism, in its strong emphasis on the need for an absolute divine inclusivity. It is, of course, extremely difficult for anyone to be totally coherent and consistent in their thinking about so profound a matter as the relation of the infinite Creator to finite creatures. Everyone is in danger of trying to impose too easily some form of logical grid upon an inherently mysterious matter. Our commonsense notions derived from everyday experience cannot be expected to serve adequately for thinking about the profundity of God's relationship to humanity or creation

14. See J. C. Polkinghorne (ed.), *The Work of Love*, SPCK/Eerdmans, 2001.

generally. Not to acknowledge this would be to end up in a position similar to that of a classical physicist who refused to recognise the idiosyncrasy of the quantum world. Neither in science nor in theology may we expect to escape entirely from paradoxical tensions, but we should only embrace paradox and mystery when they are forced upon us by the sheer undeniability of the character of what it is that is being experienced.

I certainly believe that the distant God of classical theism, existing in isolated transcendence, is a concept in need of correction by a recovered recognition of the immanent presence of the Creator to creation. However, I do not believe that this requires us to embrace the too-inclusive language of panentheism. All that is necessary is to reaffirm that creatures live in the divine presence and in the context of the activity of the living God. A concept that seems to be of value here is the distinction made in the thinking of the Eastern Church, and particularly in the writings of Maximus the Confessor and Gregory Palamas, between the divine essence (God's being, ineffable to creatures) and the divine energies (God's activity in creation). An appropriate understanding of the latter can provide a strong account of effective divine presence without endangering the distinction between creatures and their Creator.[15] I wish to consider the energies as immanently active divine operations *ad extra*.

---

15. Orthodox thinking preserves a strict distinction between the Uncreated and the created, but some forms of its discourse on the going forth of the uncreated divine energies may seem to be in danger of verging on a form of emanationism (see the complex and nuanced discussion in V. Lossky, *The Mystical Theology of the Eastern Church*, James Clarke, 1957, ch. 4). I think this tendency can arise from failing sufficiently to emphasise that *theosis* is a process that requires eschatological completion (see J. C. Polkinghorne, *The God of Hope and the End of the World*, SPCK/Yale University Press, 2002, pp. 115 and 132–36).

Acts of providence are to be understood in accordance with a recognition of the divine kenosis involved in creation, so that God is not supposed to be the agent of everything but, rather, a balance is struck between the actions of God and the actions of creatures. I have suggested, despite much theological argument to the contrary, that the Creator's self-limitation should be understood to extend even to God's condescending to act as a providential cause among causes.[16]

(2) *Trinitarian thinking.* The first Christians were monotheistic Jews who knew above all else that the God of Israel is one Lord. Yet they found that in writing and preaching about their experience of the risen and exalted Christ, they were driven to use divine language about him, even to the point of granting him the title 'Lord', which was the peculiar prerogative of the God of Israel. They also knew a divine Spirit at work in their hearts and lives, which sometimes they called the Spirit of God, sometimes the Spirit of Christ and sometimes just the (Holy) Spirit. The New Testament leaves these tensions and paradoxes unresolved, but obviously matters could not be allowed to remain for long in this intellectually unstable state. After more than three centuries of intense theological reflection and struggle, the Church formulated and embraced the doctrine of the Trinity, expressing its belief that the one true God exists in the eternal interchange of love between the three divine Persons, Father, Son and Holy Spirit.

It is important to me that Trinitarian thinking arose primarily as a response to the insistent complexity of human encounter with the reality of God experienced within the

16. J. C. Polkinghorne, 'Kenotic Creation and Divine Action'. Polkinghorne (ed.), *The Work of Love*, pp. 104-105.

growing life of the Church, and not as an act of unbridled and ungrounded metaphysical speculation. It is congenial to the thinking habits of a scientist to approach the doctrine of the Trinity 'from below' and to understand it as derived from the experience of salvation. This approach locates the theological basis of Trinitarian thinking in what the Greeks called the 'economy' (*oikonomia*), the knowledge of God that arises from the Creator's interaction with creatures. However, Catherine LaCugna, in her survey of Trinitarian theology, warns us against placing too much reliance on an approach from below:

> Theology cannot be reduced to soteriology. Nor can trinitarian theology be purely functional; trinitarian theology is not merely a summary of our experience of God. It is this, but it is also a statement, however partial, about the mystery of God's eternal being.[17]

Of course, I agree that Trinitarian thinking is not *merely* a summary of experience, any more than science is *merely* a positivistic summary of experimental data. But it is from that data that science gets its nudge in the direction of a deeper and more comprehensive understanding. Similarly, it is from the Church's experience, both soteriological and doxological, that it gets its nudge in the direction of the doctrine of the Holy Trinity. In thinking about theology I am always very conscious of the question, so natural to the scientist, 'What makes you think that this might be the case?' The rooting of understanding in experience is what I have called bottom-up thinking. Accordingly, the approach to Trinitarian thought that I find most helpful and persuasive is one that follows

17. LaCugna, *God For Us*, p. 4.

the strategy expressed in the celebrated theological aphorism called 'Rahner's Rule',[18] affirming the identity of the immanent Trinity (God in Godhead itself) with the economic Trinity (God known through creation and salvation). In other words, I rely on the belief that God's nature is truly made known through God's revelatory acts. Rahner's Rule seems to me a statement of theological realism, the assertion that what we know is a trustworthy guide to the way things are. In the case of theology, this trust is directly underwritten by the faithfulness of the God so revealed, who will not be a deceiver. It was in this spirit that I wrote that 'The proclamation of the One in Three and the Three in One is not a piece of mystical arithmetic, but a summary of data'.[19]

On reflection, it would have been better to have written 'interpretation' rather than 'summary' for, just as scientific theories are not simply read out from nature but require creative understanding of nature, so, to an even higher degree because of the veiled character of God's presence, Trinitarian thinking demands the use of creative, and indeed inspired, insight in the handling of its data. In the course of this process, second-order reflection may lead to the modification of relatively naïve first-order categorisations of that experience. An example of this happening in theology is provided by the arguments that took place in the early Church about the relationships of the Persons to each other, and the degree, if any, of subordination that might characterise these relationships. If one simply takes the testimony of the economy at face value, a subordinationist account seems to be the natural

18. K. Rahner, *The Trinity*, Burns & Oates, 1970, p. 22.
19. Polkinghorne, *Science and Christian Belief/The Faith of a Physicist*, p. 154.

conclusion. In John's gospel, Jesus says 'The Father is greater than I' (John 14:28) and there is a repeated emphasis on the *sending* of the Son by the Father, a theme also to be found in Paul (Romans 8:3). All four gospels portray Jesus as praying to his Heavenly Father. Arianism's subordination of the Son to the Father drew its support from just these types of scriptural passage. LaCugna comments that pre-Nicene Christian thinking in general concentrated primarily on the economy and this led to what she characterises as a 'patently subordinationist christology'.[20] (One sees this reflected in Irenaeus's description, cited earlier, of the Second and Third Persons as the 'hands of God'.) However, further theological reflection after the Council of Nicaea led the Church to the recognition of the unsatisfactoriness of a position that simply and naïvely equated the necessary historical phenomenology of the incarnation with the eternal divine realities. An approach from below will always have to reckon with the possibility that its understanding will need to rise above simple appropriation of the phenomena that lie at its base. (In science this corresponds to the difference between phenomenology and fundamental theory.) The essential requirement is that when theology expands and modifies its preliminary understanding, it does so for well-motivated reasons and not just in an excess of speculative exuberance.

Persuasive theoretical ideas find their support, over and above the initiating evidence, in a number of ways. One is the additional insight that can be provided, going beyond the considerations that led to the proposal being made in the first place. This overplus of interpretative success is one of the

20. LaCugna, *God For Us*, p. 23.

characteristics of deep scientific theories and it is an important factor in encouraging the belief that such theories do indeed afford us verisimilitudinous accounts of the actual nature of the physical world.[21] Trinitarian theology does not lack a similar kind of support. An unqualified monotheistic picture of God can only interpret the statement 'God is love' (1 John 4:8) as implying that from all eternity there has been within the divine nature itself an almost narcissistic state of self-regard. The Trinitarian picture of the eternal exchange of love between the divine Persons, whose communion of mutual openness constitutes the divine being, is a much more illuminating theological insight. No doubt, its articulation requires great subtlety, as theologians seek to avoid making the Trinity so 'social' that it becomes more or less a tritheistic pantheon. The concepts of perichoresis and appropriation, baffling as they may sometimes seem to the bottom-up thinker, clearly are intended to meet this need to avoid tritheism. It is beyond my limited abilities to be sure exactly how successfully this aim has been achieved by the theological proposals that have been made.

Another way in which second-order theoretical proposals can be justified is by their gaining collateral support from their consistency with other aspects of human knowledge lying outside the field of motivation for the original ideas. I suggested in an earlier chapter (p. 75) that the profound degree of relationality that science has found to be present in the fabric of the physical universe is certainly congenial to a Trinitarian way of thinking.

21. See J. C. Polkinghorne, *Beyond Science*, Cambridge University Press, 1996, ch. 2.

(3) *God and time.* One of the most significant ways in which thinkers in this present age of science are distinguished from their predecessors in earlier ages is that we are forced to take the temporal dimension of reality extremely seriously. For us, time is no longer just the index of historical events; evolutionary insight implies that it has played an essential formative role in the constitution of the present. At the end of the eighteenth century, geology was the first science to realise this, followed by biology in the nineteenth century. However, the resistance of the physicists to accepting the indispensability of fully temporal thinking endured well into the twentieth century. When Einstein formulated the first truly scientific cosmology in 1917, he supposed that he had to find a static model for the universe. It was only Edwin Hubble's discovery of the recession of the galaxies that led the physical cosmologists to take seriously the solutions of Einstein's equations, discovered by the Russian meteorologist Alexander Friedman and the Belgian priest Georges Lemaître, which described a temporally changing and expanding cosmos. It is impossible today to think about created reality without acknowledging its evolutionary character and its radical temporality.

It is a fundamental theological belief that God's knowledge is totally true and wholly comprehensive. God knows things as they really are. If that is the case, and if God's creation is intrinsically temporal, surely the Creator must know it in its temporality. In other words, God will not simply know that events are successive but God will know them according to their nature, that is to say, *in their succession.* Such a thought runs totally counter to the concepts of classical theology, where God was believed to know history ahistorically, 'looking down', so to speak, from an atemporal viewpoint onto

the whole of history laid out before the divine gaze, perceiving it *totum simul*, all at once, to use Boethius's famous phrase.
I raised in an earlier chapter (p. 81) the question of whether
a universe that we now regard as a world of true becoming,
rather than a world of static being, could properly be known
in this way.

An alternative way of thinking is offered to us by process theology, which proposes a dipolar concept of God's
relationship to time. It suggests that there is in the divine
being both an atemporal eternal pole (which it calls the divine
primordial nature) and a temporal pole (the divine conse-
quent nature). It is not necessary to accept the detailed scheme
of process metaphysics, within which A. N. Whitehead and
Charles Hartshorne embedded divine dipolarity,[22] in order to
see that here is an idea capable of fruitful theological development. In fact, much twentieth-century theological discussion,
within a variety of traditions, took God's engagement with
time with great seriousness. Divine temporal dipolarity has
been an important component in the thinking of the scientist-
theologians.[23] It does not involve some abstract conflation of
time and eternity, but rather a mutual complementarity between the unchangingly steadfast and the providentially re-
sponsive. Motivation for adopting such a point of view does
not arise solely from a science-influenced theology of nature,
but is also to be found in the biblical resources themselves. The
Bible is very much concerned with God's engagement with
temporal process—in the history of Israel and, above all, in the
historical episode of the incarnation, spanning the time inter-

22. A. N. Whitehead, *Process and Reality*, Free Press, 1978; C. Hartshorne,
The Divine Reality, Yale University Press, 1948.
23. J. C. Polkinghorne, *Scientists as Theologians*, SPCK, 1996, p. 41.

val from a birth under Augustus to a death under Tiberius. We have seen that the strongly anthropomorphic language of the Hebrew Bible is unrestrained in its expression of divine temporality, even to the extent of portraying God as changing God's mind. Following the apostasy of the golden calf, the Lord says that divine wrath will consume Israel, but after pleading by Moses 'the Lord changed his mind about the disaster he planned to bring on his people' (Exodus 32:14). In a later incident, imminent catastrophe is decreed as a punishment for Ahab's complicity in the judicial murder of Naboth, but on perceiving his penitence, God decides to postpone these consequences till after the king's death (1 Kings 21:28–29). The sequence of the 'days' of creation in Genesis 1 represents a God who acts sequentially and not timelessly (see also Proverbs 8:22), something that was a problem for Augustine's Platonic way of thinking, which saw time as coming into being as a consequence of a single atemporal creative act.

A number of comments need to be made about the theological consequences of accepting the insight of a divine temporal dimension. The first, of course, is to emphasise the essential importance of the divine atemporal pole. There has to be an eternal and unchanging dimension to the divine reality, in order for there to be a basis for the steadfast love that is God's nature, and for the certain hope for creation that is God's promise. God is certainly not in thrall to time in the way that all creatures are in thrall. Creation as we know it came into being at the singular moment of the Big Bang, some fourteen billion years ago. As Augustine understood fifteen centuries before Einstein, the beginning of the created universe was the beginning of time. But it certainly was not the beginning of God, though surely it was the occasion for the

Creator graciously to embrace a temporal relationship to creation. This remark reminds us, in a particularly strong way, that divine temporality implies the possibility of divine mutability. This does not arise from a subversion of the steadfast being of God, but through the conformation of the divine relationship with creation to the actual, and temporally changing, character of that creation. One must surely suppose that the Creator related to the universe in one way when it was a ten-billionth of a second old and consisted simply of an energetic soup of quarks, gluons and leptons, and in a different way today, fourteen billion years later, when the universe is the home of saints and sinners.

Classical theology was fiercely insistent on the immutability of God. Partly that was because it rightly sought to exclude any thought of the possibility of the external magical manipulation of God by creatures, though it failed to conceive of the idea of God's internal willingness to relate in different ways to creation at different stages of its history. Classical theology was always in bondage to the Greek idea that perfection lies in the attainment and maintaining of a static maximal state, from which to depart in any respect would inevitably be to decline. According to this view, a perfect being had to be immutable, because there was nowhere else to go. Behind this whole way of thinking lay the instinctive Greek conviction that the eternal is always better than the temporal. With our grasp of the fundamental fruitfulness of evolving temporality, it is possible for us to embrace a contrastingly dynamic view of perfection. God's utter perfection lies in the total appropriateness at all times of the Creator's relationship with creation, whether that creation is a quark soup or the home of humanity.

God's acceptance of a temporal pole within divinity is another instance of the kenosis of the Creator in the act of creation, as the One who is eternally steadfast enters into a relationship with creatures that is conditioned by time. In turn this leads to a kenosis of divine knowledge, a self-limitation of God's way of knowing. If the universe is truly a world of becoming, the future is not up there, waiting for us to arrive, for we help to make it as we go along. It would seem that even God does not have that future available for perusal beforehand. All theists would want to affirm divine ominiscience. The classical theist does this in absolute terms, asserting that God knows atemporally the whole of history 'at once'. In the scheme of process theology, God is the reservoir of all memory of the past, but does not know the future, because of a metaphysical *necessity* that the divine nature be limited in this way. In the kind of dipolar theism that I am seeking to espouse, God is understood to have *chosen* to possess only a current omniscience, temporally indexed. God knows now all that can be known now, but God does not yet know all that will eventually become knowable. Of course, God will not be caught out by the movements of history into its future, in the way that human beings are so often caught out. The Creator can see how creation is developing, but God does not yet have a detailed acquaintance with how that development will turn out as the future unfolds. This does not negate ultimate divine sovereignty, for we may suppose that God can bring about determinate ends through contingent paths. Think of William James's picture of the Grandmaster of cosmic chess, who will win the game whatever moves the creaturely opponent may make.

Understanding divine omniscience in this temporal way seems compatible with the way that prophecy works in the Bible. The prophet's message is expressed in general terms and its consequences are subject to revision if human free response takes history in a new direction. We can see this in the way that the Lord speaks conditionally to Ezekiel about the warnings and judgements he is to convey to Judah (Ezekiel 33). The fact that God could warn that nation through Jeremiah that Egypt would not deliver Jerusalem from the invading Babylonians (Jeremiah 37:3-16) does not imply that God saw beforehand exactly when, and through which individuals, the Temple would be put to the torch.

Much recent theological thinking has acknowledged that God's appropriate relationship with creation includes divine suffering in compassionate solidarity with the travail of creatures.[24] This divine passibility implies an openness to mutability, for 'suffering in the widest sense means the capacity to be acted upon, to be changed, to be moved, transformed by the action of, or in relation to, another'.[25] Here is another motivation for taking God's engagement with time with the utmost seriousness.

Finally we must address the question of what time is to be associated with divine temporality. Relativity theory abolished Newton's universal time, replacing it by a multitude of times associated with localised observers, whose judgements of the simultaneity of distant events are retrospective assessments that depend upon the observer's state of motion. God, of course, is not a localised observer, but is omnipresent to cre-

24. A particularly helpful account of this idea is given in P. Fiddes, *The Creative Suffering of God*, Oxford University Press, 1988.
25. D. D. Williams, quoted by Fiddes, ibid., p. 50.

ation.[26] For the Creator there are no 'distant events'. Whatever time is associated with the temporal pole in God, God will know every event in creation *as and when it happens.* Since this is so, it does not seem to me a very critical question for theology as to which time is God's time. In fact, however, there is a particular choice that one might suppose the Creator to have been likely to have made. When we consider the physical universe as a single entity, there is a natural frame of reference that has a preferred status. This is the frame that is at rest with respect to the cosmic background radiation (that is to say, at rest overall with respect to the universe taken as a whole). The time associated with this frame is the time that cosmologists use to define the age of the universe and to date its earlier epochs. It would not be surprising if the Creator had made the same choice. If that is the case, it would illustrate that when we think of the universe as a creation we should not hesitate to take with great seriousness the details of its particularity (in this case its large-scale homogeneity, which permits the definition of cosmic time).

(4) *Divine complexity.* One of the most difficult assertions made by classical theology, but one which is a keystone in Thomas Aquinas's impressive intellectual construction of the *Summa Theologiae*, is that of divine simplicity (Ia, q. 3). In Thomistic thinking, the divine nature is single in the strongest sense, not to be differentiated into elements of any kind, such as divine knowledge or divine will, lest they should seem to be logically prior to God. It is not supposed to be the case that the divine justice can be assessed against some free-standing

---

26. Spatial omnipresence and divine temporality are compatible concepts, since relativity theory distinguishes between spacelike and timelike relationships. History remains different from geography.

concept of what is just for, in Thomas's view, God does not participate in justice, but is justice itself. There are notorious difficulties in following through these ideas. If God knows the evil that there is in creation but does not will it, surely the divine knowledge and the divine will are to be distinguished from each other. God's justice is not the arbitrary decree of a celestial despot, but is in absolute conformity to what is right.

The topics we have already discussed in this chapter, if one follows the lines I have taken, put divine simplicity under further strain. However subtly the discussion may be nuanced, Trinitarian thinking surely indicates a degree of complexity existing eternally within the divine nature. Resistance by the Western Church to acknowledging this has resulted in what Jürgen Moltmann has called a 'Christian monotheism', which he believes fails to be truly Trinitarian in its thinking.[27] If the idea of a divine dipolarity of eternity–temporality is taken seriously, it gives further grounds for believing there to be divine complexity contained within the one being of God.

Divine polarities also seem to Keith Ward to be necessary for an adequate understanding of the theology of creation. He asks, if God is purely eternal, how can God's action give rise to anything that is intrinsically temporal? Ward also emphasises what he sees as the related need for a polarity of necessity and contingency within the divine nature, for he believes that it is equally difficult to see how a purely necessary being can be the ground of the existence of a contingent creation. In Ward's view, 'if the world is to be contingent, and

27. J. Moltmann, *The Trinity and the Kingdom of God*, SCM Press, 1981, pp. 129–50.

man really free, contingency and mutability must exist within God himself'.[28] This leads him to the conclusion that 'The general solution is to say that there is just one individual, possessing both a set of necessary properties and a set of contingent properties'.[29] A strongly unqualified divine simplicity is, therefore, to be rejected. Ward believes that 'the whole Doctrine of divine simplicity arises from a misinterpretation of the truth that God's properties are not divisible into parts, that all his properties are interconnected. God's properties are necessarily connected in a reciprocal determining whole and are not just contingently or fortuitously related'. He goes on to say that he can see 'no a priori reason why the Divine being should not be internally complex, each part depending essentially on the unity of the whole'.[30] Divine unity is an indivisibly integrated complexity. Even a more traditional theologian such as Catherine LaCugna can express what seems to be a degree of implicit sympathy with this point of view. She states that Rahner's Rule 'reaffirms the basic unity of the theological enterprise, and removes once and for all the compartmentalism of theological themes and separate entities . . . there is only one God, one self-communication, one triune mystery of love and communion, *which has both eternal and temporal modalities*'.[31]

The idea of divine complexity-in-unity gains some analogical support from what can be understood about human personhood. Human beings are neither schizophrenically fragmented nor lacking in internal complexity. From Augus-

28. Ward, *Rational Theology and the Creativity of God*, pp. 80–81.
29. Ibid., p. 165.
30. Ibid., p. 64.
31. LaCugna, *God For Us*, p. 231, my italics.

tine's differentiation of memory, intellect and will within the single person (which, of course, he used as an analogy for the Trinity), to the varied and partially conflicting deliverances of modern depth psychology concerning the conscious and unconscious mind, it has been recognised that human beings are much more complex than their conscious experience of the rational ego would have seemed to suggest. Yet each person also possesses an individual integrity. If there is this subtle structure to the finite human person, it does not appear at all surprising that there is complexity-within-unity in the case of infinite deity also.

(5) *Incarnational thinking.* The incarnation is central to orthodox Christian theology. At the heart of Christian belief is the mysterious and exciting idea that God has acted to make the divine nature known to us in the clearest and most accessible way, by living the human life of the Word made flesh in Jesus of Nazareth. For Christians it is also truly the case that God is 'the fellow sufferer who understands', through divine presence in the cross and passion of Jesus. A profound exposition of this understanding has been given in Moltmann's Trinitarian interpretation of Calvary as the event in which 'the Son suffers dying, the Father suffers the death of the Son'.[32] In fact, Moltmann claims that Trinitarian thinking centres on the cross, so that he can go so far as to say 'The content of the doctrine of the Trinity is the real cross of Christ himself. The form of the crucified Christ is the Trinity'.[33]

I believe in the truth of the incarnation. This is not the place to rehearse all the motivations that support me in that

32. Moltmann, *The Trinity and the Kingdom of God*, p. 243.
33. Ibid., p. 246.

belief,[34] but I do want to emphasise the importance of soteriology as the key to Christology. What Christ has done reveals who he is. The New Testament resounds with the conviction of its authors that they have found in the risen Christ a source of new and transforming life in this world, which will continue beyond death in the everlasting life of the world to come. If we are to make sense of this saving encounter, it is necessary that Jesus be fully human, totally immersed in the life of humanity and so totally relevant to every aspect of our lives—he assumes all to redeem all, as Gregory Nazianzus stated so well—yet he must also be truly divine so that it is the invincible life of God to which we are given access through him. Christ is to be recognised precisely as the One who, in the two natures, human and divine, constitutes the bridge between the infinite life of the Creator and the finite lives of creatures.

So paradoxical and exciting a belief is forced upon the Christian by salvific experience, but its mysterious character is such that it will not yield to being completely explained by human reasoning. On the one hand, we need to be realistically modest about what Christological thinking can be expected to achieve while, on the other hand, we must not succumb to intellectual laziness in refusing to attempt the task at all. One question to be addressed is how one is to understand the continuing providential governance of the universe during the episode of the earthly life of the incarnate God. No one supposes the throne of heaven to have been vacant during that time. Trinitarian thinking comes to our aid here, for it is only the Second Person of the Trinity who is believed to

34. See Polkinghorne, *Science and Christian Belief/The Faith of a Physicist*, ch. 7.

have 'humbled himself and become obedient to the point of death—even death on a cross' (Philippians 2:8). Even so, there has been a current of Christian thought, particularly within Calvinism, that has wanted to see the Word as continuing to participate in the governing of creation (an idea some Lutherans nicknamed the *extra Calvinisticum*).

I feel sympathy with this Calvinist intuition. With some trepidation, and certainly with considerable tentativeness, I venture to suggest that the idea of divine eternal-temporal dipolarity might be of significant help here. Would it not be sufficient for Christian theology to suppose that it was *the temporal pole* of the Second Person that became incarnate in Jesus of Nazareth, while the eternal pole continued its timeless participation in the divine essence and governance?

There seem to me to be two reasons for considering this possibility. One is that the incarnation represents the utmost conceivable engagement of the divine with the temporal, and so it appears wholly appropriate that it is the temporal pole which participates in this sharing with creation. The second reason is that I believe the idea passes the soteriological test for an adequate Christology. In Chapter 6 I shall seek to explore the implications for eschatology of what I believe to be the essentially temporal nature of human beings. This will lead to the conviction that our destiny beyond death is not to enter into some timeless state of eternity, but to live an everlastingly redeemed life within the 'time' of the new creation. This destiny will indeed be *theosis*, a sharing in the life of God, but it will not be human participation in the ineffable life of the eternal divine pole. Rather, it will be an unending exploration of the riches of the temporal pole of deity, made accessible to us in Christ. To bring about this salvific experience it seems necessary only for the temporal pole of God to be united with the temporal life of creatures, in the 'bridging operation' of the incarnation. Finally, if one follows Moltmann in his Trinitarian account of Calvary, one might suppose it to be the temporal poles of all three divine Persons that were involved in that historical event. Yet the intimate complementarity of temporal and eternal within the unity of God also enables one rightly to speak of the Lamb slain from the foundation of the world (see Revelation 13:8) and to speak of continuing divine participation in the travail of creation.

We have reached a point at which it is perhaps best for the theological reflections of a bottom-up thinker to come to a close. Even in the hands of lifetime professionals, there are limits to the completeness of understanding that can be achieved through the human enterprise of theology. Yet that recognition should not deter anyone from attempting the task of theological thinking, nor encourage them to settle prematurely for some relatively undemanding form of understanding. It is not naïve simplicity that will be persuasive in metaphysical discussion. Accounts that are truly convincing may be expected to have a complexity and richness about them that place demands on our thinking. One sees that this is so in science too. Bohr's model of the atom depended on one simplifying assumption (circular orbits) and one simple rule added to Newtonian physics (the quantisation of angular momentum). It certainly offered insight, but it was too 'thin' to give a lasting picture. Eventually one needed the 'thick' theory of quantum mechanics, with its much greater complexity and the counterintuitive character of its thought. It is scarcely surprising that in theology also it is only a thick account that proves even partially adequate to the exploration of the inexhaustible riches of the Trinitarian God, the Ground of our existence and the Source of our everlasting hope.

CHAPTER FIVE

The Eucharist: Liturgy-assisted Logic

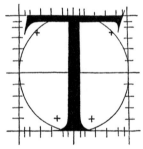HE bottom-up thinker seeks to move from experience to understanding. In science this transition occurs under the pressure of the questions that the experimentalists pose to the theorists, often pushing them to the discovery of ideas that they would not have been able to find, or even to imagine, without the necessary nudge of nature. Niels Bohr once said that the world is not only stranger than we thought, it is stranger than we could have thought. One might say that the reasoning of science advances by a process of universe-assisted logic.

I believe that there is an analogous movement in theology, where the faith that is seeking understanding receives its impetus from religious experience.[1] This experience is of

1. See J. C. Polkinghorne, *Reason and Reality*, SPCK/Trinity Press International, 1991, ch. 1; *Belief in God in an Age of Science*, Yale University Press, 1998, pp. 116–22. Defence of the evidential value of religious experience is given in R. Swin-

a variety of kinds. A substantial part, of course, is vicarious, deriving from the acceptance of the accounts of the foundational events and insights recorded in scripture. A further part also comes to us externally, from the testimony of outstanding religious figures, the kind of 'pattern-setters' that so interested William James in his important Gifford Lectures, *The Varieties of Religious Experience*.[2] Every theologian must make use of these resources, which together constitute a canonical body of classical Christian literature.[3] But, for a living encounter between faith and understanding, there must also be an internal resource. The ancient meaning of the word 'theologian' was not narrowly academic, but referred to someone who not only thought about God but was also a person of prayer, someone whose own religious experience was an indispensable part of their theological life. I wish to number myself as a humble member of that company, though I must confess that I am not a case that would have been of any interest to William James. My Christian life is central to who I am, but I have to acknowledge that mine is a rather humdrum kind of spirituality. For me, Christian practice centres on a certain degree of faithfulness in prayer, worship and service. A particularly important part of this experience is located in my regular participation, week by week, in the Eucharistic celebration of the Church. Sometimes I have the privilege of presiding at the altar on behalf of the gathered community of the faithful, and sometimes I am simply a member of the congregation. Whatever

---

burne, *The Existence of God*, Oxford University Press, 1979, ch. 13; W. P. Alston, *Perceiving God*, Cornell University Press, 1991.

2. W. James, *The Varieties of Religious Experience*, Longmans, Green & Co., 1902.

3. See D. Tracy, *The Analogical Imagination*, SCM Press, 1981.

the role, that regular sharing in Holy Communion is an indispensable element in my Christian life. For me, theological thinking proceeds by a kind of 'liturgy-assisted logic'.[4] I want in this chapter to explore a little of what that might mean. Science as such will be relevant to this task only to the extent that it encourages reliance on interpreted experience as the route to truth.

The first thing that I want to say is that I understand this life of regular Eucharistic practice to be a fulfilment of the Lord's command to do this in remembrance (*anamnesis*) of him. Of course, I know that to make this assertion immediately raises tricky scholarly questions.[5] The oldest account we have of the institution of the Lord's Supper is given by Paul in his first letter to the Corinthians (11:23–26), which conveys the dominical command for remembrance in relation to both the bread and the cup. In Luke (22:19–20), the explicit command is given in relation to the bread only. At least that is so in all the early Greek manuscripts except for Codex Bezae, which gives a severely truncated account, omitting the command altogether. In spite of the weight of manuscript testimony supporting the longer reading, many scholars have favoured attributing authenticity to the shorter text, principally on the grounds of the text-critical injunction generally to prefer the shorter reading, since it is easier to imagine how material might subsequently be added than to imagine how it would come to have been omitted. In Mark (14:22–24), and in the essentially parallel passage in Matthew (26:26–28), there is no command to make remembrance. All the accounts have

---

4. Polkinghorne, *Reason and Reality*, p. 19.
5. See J. Jeremias, *The Eucharistic Words of Jesus*, SCM Press, 1966.

Jesus associating the bread with his body and the cup with his blood, the latter linked in some way with covenant.

The assessment of this material is complicated by the fact that the 'breaking of bread' is testified to have been a regular practice of the Christian community from the very first (Acts 2:46) and so the way in which the words of Jesus were recalled and conveyed to others would have been subject to primitive liturgical influence and shaping from the beginning. From a theological point of view, however, I do not think one has to attain absolute certainty about historical detail in order to be able to understand the continuing celebration of the Eucharist over the ensuing centuries as being the Church's fulfilment of a command from its Lord. It seems absolutely clear to me that the Lord's Supper derives from the words and actions of Jesus himself on the night of his betrayal. One cannot suppose that the identification of the wine with the blood, so naturally repulsive an idea to normal Jewish thinking (cf. Genesis 9:4 and Acts 15:20b), could have arisen and been universally accepted in the first generation of Christians unless it had been known to have dominical authority. If this had not been the case, there would surely have been evidence of dissent in the early Church about the issue, comparable to that which did indeed arise about the status of circumcision. When one reads Paul prefacing his account of the sacrament by saying that he received it 'from the Lord', there are a variety of ways in which this phrase might be understood. One would be that Paul received a report handed on from those who had known Jesus in the flesh, giving a direct historical reminiscence of what he had actually said at the Last Supper. This seems to me the most probable meaning, but there are other possibilities. A much more speculative idea would be that during the period of the resurrection appearances the risen Christ conveyed more teaching to the apostles than the small amount recorded in the gospels and that the command to make remembrance was part of this 'secret teaching'. Or one could believe that the command to continue the remembrance was conveyed by the inspiration of the Holy Spirit, the Spirit of Christ, at work in the primitive Church in the immediate aftermath of Easter and Pentecost. Whichever may be the right way to think about the matter, if we believe in the continuing Trinitarian activity of God in the Church, it is indeed 'from the Lord' that we receive the command to do this in remembrance of him, and it is our duty and our joy to be obedient to that injunction.

The accounts of the Last Supper are not the only scriptural loci to which we can look in thinking about Holy Communion. It is well known, and one has to say very perplexing, that in its version of the events of that fateful Thursday evening, John's gospel replaces the account of the institution with the story of the foot-washing. The author of the fourth gospel chose to give what is clearly very strong Eucharistic teaching in the different context of the feeding of the multitude (John 6:41–65). The stories of the miraculous feeding are given in all four gospels and their relationship to Holy Communion has often been remarked upon. In particular, Dom Gregory Dix has emphasised that the fourfold action of Jesus at the feeding, in taking, blessing, breaking and giving (actions also performed at the Last Supper), mirrors the pattern that Dix regarded as being the shape of the basic fourfold action of the Mass, enacted in obedience to the Lord's command.[6]

6. G. Dix, *The Shape of the Liturgy*, Dacre Press/Adam & Charles Black, 1945.

It is also clear that the story of the disciples on the road to Emmaus, with its veiled encounter with the one who makes their hearts burn within them through the exposition of scripture and which reaches its climax in the explicit recognition of the risen Lord at the moment of his breaking bread (which is also the moment of his disappearance from their sight), was treasured by the Church at least partly for its Eucharistic associations. Once again we encounter the pattern of the four-fold action (Luke 24:30). The story also endorses the common Christian practice that the celebration of the Eucharist does not consist only in the sacramental action itself, but also includes the ministry of the word in which scripture is read and expounded.

These considerations all converge in giving us scriptural warrant to regard Holy Communion as indeed a dominical sacrament, instituted by Christ as a focused and covenanted occasion of particular divine presence with the faithful. Of course, God is never absent from creation, but sacraments are specific events in which the veil that covers the divine presence is thinned for the believer. In this way, the Lord's Supper is quite different from an ordinary meal enjoyed with overtones of piety.

Outwardly, in its appearance, what happens is simplicity itself. People pray and sing; words of Jesus are recalled; everyone eats a small piece of bread or a wafer, and takes a sip of wine; there is a concluding blessing and then all go home. What matters, however, is the meaning of what is happening, the interpretation of what is going on in the sacrament. A scientist knows that happenings which seem rather trivial in their surface character (a messy bubble chamber photograph: X-ray photographs of a crystal, and so on) may be profoundly significant when they are interpreted (for example, a further step on the way to the discovery of a new level in the structure of matter; the helical nature of DNA).

A good deal of the Church's understanding of what is going on in Holy Communion is conveyed by the prayers that are said at its celebration, and in particular the great prayer of thanksgiving (the *anaphora*), which lies at its heart and which incorporates the recollection of the instituting words of Jesus. That is why I have called the theological understanding that flows from sacramental practice *liturgy*-assisted logic. Different Christian traditions weigh the importance of a canonical liturgical setting in different ways. A Reformed theologian like Michael Welker, in his extremely helpful and insightful book *What Happens at Holy Communion?*,[7] pays no detailed attention to the varied forms of liturgical expression in use in the Church. On the other hand, a Roman Catholic theologian like Father Louis Bouyer can say of the *anaphora* that it is the 'core' of the Eucharist and he devotes the greater part of his book on the sacrament to a detailed scholarly review of the historical development of liturgical traditions.[8] In characteristic Anglican fashion, I find myself somewhere in the middle, concerned with the shape of the liturgy but in a way that is controlled by the wider considerations of sacramental theology. The actual form of the liturgy is certainly important for Anglicans, not least because it provides the basis for the unity of our ecclesiastical Communion. Anglicanism has never had a paradigm theological figure of the calibre of Aquinas, Luther or Calvin, or a detailed, explicit confessional basis such as that provided

---

7. M. Welker, *What Happens at Holy Communion?*, Eerdmans, 2000. I make considerable use of the ideas of this book in what follows.

8. L. Bouyer, *Eucharist*, University of Notre Dame Press, 1968.

for Roman Catholics by the teachings of the magisterium, for Lutherans by the Augsburg Confession, and for Scottish Presbyterians by the Westminster Confession. The Thirty-Nine Articles are too unsystematic to play that role and today they are recognised as being very timebound in many of the concerns that they address. Anglican ecclesiology can be summed up by the adaptation of a familiar religious slogan, for it is our belief that the church that prays together, stays together. If you can use the words of the Liturgy with sincerity, you are an Anglican, without the authorities having to cross-examine you about detailed and particular matters of the interpretation of these words. Here is the source of Anglican comprehensiveness—or Anglican wooliness, according to how you view the matter.

Although the Book of Common Prayer of 1662 no longer fully functions as the live basis of pan-Anglicanism, the various modern provincial liturgies still have a sufficient degree of family resemblance for Anglicans to feel at home wherever they worship throughout the Communion. And liturgical developments of the past fifty years in many other Christian churches have displayed a significant measure of mutual convergence in the forms of service. I believe that this can be taken as an encouraging sign that there is less scandalous division and controversy among us today about the nature of the sacrament than in the past. One has only to recite the alternative ways of naming the sacrament—Mass, Eucharist, Holy Communion, Lord's Supper—to recall the intensity of past disagreements. Of course, the variety of names still persists and the differences of perspective that it represents have not disappeared, but it is easier than it was to recognise that all Christians are seeking to obey the Lord's command to do this in remembrance of him, and also easier to believe that each tradition has preserved insights in which it is necessary for all to share if we are to gain a deeper understanding of the Holy Mysteries. Real convergence is taking place among us.

There are certain sacramental understandings that would be widely agreed. First, and of very great importance, is the recognition that Holy Communion is celebrated *in the presence of the risen Christ.* In modern liturgies, the prayer of thanksgiving often begins with the president affirming 'The Lord is here', to which the congregation reply 'His Spirit is with us'. Over history there have been many and bitter arguments about whether and how there is a real presence of Christ in the celebration of the Eucharist. Positions were taken ranging from a Zwinglian symbolic understanding of the sacrament, which saw it as acting simply as a sign to prompt the spiritual reception of Christ by the individual faithful believer, to a Tridentine insistence on the transubstantiated nature of the elements, which made them the carriers of the adorable and persisting reality of the body and blood of the Lord. While these differences have not totally disappeared, there is an increasing recognition that the heart of the matter lies with the reality that Jesus Christ is present in his risen and exalted personhood in the whole action of the Eucharist, the sacrament at which Christ is both the giver and the gift. Here is the core of our common Eucharistic experience, the phenomenon to which Christians wish to bear their grateful testimony, even amid a continuing and not wholly compatible variety of theological proposals for how that phenomenon is best to be understood.

It seems to me that this acknowledgement of the Lord's presence should serve as a sufficient basis for the possibility of the Eucharistic hospitality that so many of us yearn to see coming to be shared among more and more Christian people. Such a sharing certainly requires the foundation of a commonality of intent in order to sustain it, but this is surely adequately provided by the desire to meet in the presence of the risen Lord, and by the common conviction that he has promised to honour the covenant of just such a sacramental encounter. Sharing does not seem to require that all present subscribe to a particular and detailed dogmatic understanding of sacramental theology, and it would appear doubtful whether this would in fact be the case for any body of worshippers in any church community. The situation is analogous to that which we considered when scripture was under discussion (p. 43). It is entirely possible to recognise others who share one's acknowledgement of the authority of the Bible, without all being agreed about exactly how that authority is best defined. In a similar way, one can discern a Christian comradeship with those who reverently approach the sacrament affirming their expectation of a promised meeting with the real presence of the risen Christ, however they would express their understanding of that reality.

While the divine context of Holy Communion is the presence of Christ, there is also simultaneously another context, human in its nature and quite different in its character. It goes back to the Last Supper itself, for it is *the context of threat and betrayal.* This insight is not widely recognised, but its significance has recently been powerfully emphasised by Welker. He reminds us that 'Holy Communion is instituted in the night when Jesus Christ is betrayed and handed over to the powers of this world. It continually bears the impact of this background'.[9] Welker draws a contrast with the Jewish Passover meal. Its foundational context is also that of a threatened community, in this case held in slavery by the powers of Egypt. At the Passover, though the setting is one of affliction and distress, the threat of danger comes only from outside the Israelite community itself. In the case of the Last Supper there is also external threat, stemming from the activity of the Jewish leaders and the Roman authorities (the religious and civil powers of the day), as they begin to mobilise their forces for the removal of the dangerous embarrassment, and threat to the continuing stability of society on their terms, that is posed by Jesus of Nazareth. But here there is also internal danger and betrayal. Judas, the one who will hand Jesus over to suffering and death, is sitting there at the table. The scriptural record gives us absolutely no reason to suppose that he did not also receive a piece of bread and take a sip of wine after Jesus had spoken those strange words about his body and his blood. Peter is also there, the one who in a few hours' time is to deny point-blank and with anger that he has even heard of Jesus. The other disciples are also part of the company—those who will run for their lives as soon as danger threatens. Even at the table, they were pretty jumpy, nervously saying 'Surely, not I?' (Mark 14:19) when Jesus straight out predicted that one of them would betray him.

The ambiguous character of those gathered for the Eucharist has continued in the Church. That was why Paul had to issue a stern warning to the Corinthians that 'Whoever eats the bread or drinks the cup of the Lord in an unworthy manner will be answerable for the body and blood of the Lord'

9. Welker, *What Happens at Holy Communion?*, p. 43.

(1 Corinthians 11:27). These insights place the Church in a delicate and difficult situation. The solemn nature of Holy Communion means that it cannot be an occasion on which anything goes. Yet, consideration of the company present at the Last Supper does not encourage the idea of 'fencing the table', excluding from the sacrament anyone whose position seems to be at all dubious when subjected to scrutiny. Even less are we encouraged to accept the notion that the sacrament offers a handy means of ecclesiastical discipline, so that excommunication is a useful sanction to employ against the heretical and the recalcitrant. One cannot help recalling all those dinner parties in the gospels at which Jesus is present and where the guests are questionable people like tax collectors and other sinners. It is clear that Jesus' scandalous willingness to eat with such persons, without insisting on prior public acts of repentance, gave considerable offence to his respectable religious contemporaries (Mark 2:15-17). Yet it is also impossible to believe that these meals were occasions of moral laissez-faire, which had no transforming effect on those who were participants. They were surely times when people began to be drawn out of the old life into a new and ennobled way of life, committed to following the way of Jesus. When Jesus invited himself to dinner with that old scoundrel Zacchaeus, not only did the crowd grumble but Zacchaeus's life was changed as he made fourfold restitution for his cheating and gave half of his possessions to the poor (Luke 19:1-10).

The Eucharist is too profound an occasion to be free of discipline altogether, but the reconciling and accepting nature of the sacrament is such that it must never be allowed to become the private preserve of an ecclesiastical in-group. A churchwarden is right to restrain a tipsy reveller who has wandered idly into church at the Midnight Mass on Christmas Eve, but the sacrament is not the possession of the perfect, for it is the place where sinners can find welcome and forgiveness.

One of the most important aspects of Eucharistic theology, recovered and widely acknowledged today, is the insight that considers the sacrament in its entirety, seeing it as taking place in *the total action of the gathered Christian community*. A great deal of medieval and Reformation thinking on the Eucharist was minutely focused, concentrating on a so-called 'moment of consecration' and, in consequence, centring its attention on the nature and status of the Eucharistic elements following that consecration, and especially discussing what resulted from the repetition of Christ's words of institution. Modern thinking, including much Roman Catholic thinking in the wake of Vatican II, is much more holistic in its approach. It is the whole action of the Supper that is celebrated in the presence of its risen and exalted Lord. Many important understandings follow from adopting this holistic approach.

One is a welcome relaxation of tension about precisely what happens exactly when and exactly where. The essential affirmation that needs to be made is seen to be that the Lord is present in the whole sacramental action; the body and blood of Christ are received in the course of the full unfolding of the Eucharist. The central prayer of the liturgy is indeed a prayer of thanksgiving, and not simply a formula for successful consecration. The attempt to affirm too much specificity, and to subject Holy Communion to too detailed an analysis, has often had the result of distorting sacramental theology. In the case of Roman Catholic thinking this has tended to carry the danger of too great a degree of reification, focusing on the consecrated elements; in the case of Protestant thinking this has tended to carry the danger of too great a degree of spiritualisation, focusing on the feelings of the individual worshipper. It is the total action of the whole gathered people of God, present with the gifts that they have laid upon the holy table, that constitutes the valid sacrament. The Holy Mystery is a sacramental process through which the faithful worshipper receives the body and blood of our Lord Jesus Christ, and not a single identifiable and isolatable moment within the service.

In this Eucharistic action there is a special role for those who have been authorised by the Christian community to act on its behalf, who therefore have the privilege of presiding at the celebration of the sacrament and at the administration of communion. In my own Christian tradition, this privilege attaches to the presbyterial office, in which the ordained priest is authorised to act on behalf of the priestly company of all believers. I am grateful to be a member of that historic order, but I do not suppose that those ministers whose authorisation is conveyed to them by a different process are incapable of celebrating a valid Eucharist. Certainly, all Christian traditions must have some form of authorisation of their sacramental ministers, since the privilege conveyed is not one that can be taken up by anyone lightly, on the spur of the moment and solely on their own initiative. I also believe that none of us has the fullest possible degree of sacramental authorisation, since that could only come from the whole Church of God and our present regrettable divisions withhold that total degree of authorisation from everyone. In consequence, when two Christian bodies are reconciled to each other and fully accept each other's ministries, I feel no difficulty about there being an appropriate service involving the mutual laying on of hands as a process symbolising and effecting the fusion of two previously separated Eucharistic communities. This would not deny what each had previously received, but it would enhance the fulness of the continuing gift.

Although there will be a presiding minister at Holy Communion, acting visibly on behalf of Christ, the true host at the sacramental meal, the fact that the Eucharist is the action of the whole gathered community implies that the rest of the members of the congregation are not there in a subsidiary or spectatorial role. Their participation is as important and essential as that of the celebrant, precisely because all share in the fundamental priesthood of all believers (1 Peter 2:9–10). In my own church, canon law forbids the celebration of Communion by a priest alone.

The communal character of the sacrament means that *the communion is with each other as well as with God.* In the gospel of Matthew, Jesus bids us be reconciled with each other before we bring our gifts to the altar (5:23). There was an old tradition of Anglican piety that counselled people to go to the sacrament and to return from it without speaking to anyone. The intention was reverent but its form was distorted. Today, fortunately, this practice is largely forgotten. In fact, our problem lies in altogether the opposite direction. Often, the way that modern liturgy is celebrated can appear to encourage a kind of chatty sacramental ambience that seems in danger of taking the presence of Christ on too easy and familiar terms, laying great stress on the horizontal relationships between the worshippers at the expense of an adequate recognition of the vertical relationship with the One who is worshipped. I have been in churches in which the high point of the service has appeared to be the exchange of the Peace, resulting in prolonged excursions round the building to greet and hug friends. The practice has seemed to imply that this is what Holy Communion is really about, with the subsequent reception of the sacrament almost being something of an anti-climax.

What role then do *the gifts of bread and wine* have in all this? They are surely of great importance, but not in a manner that is detachable from the totality of what is going on. It seems to me of great significance that bread and wine are not only gifts of created nature in that they derive from wheat and grapes, but are also the products of human labour. In liturgical words that are often used at the Offertory, the gifts are 'what earth has given and human hands have made'. They represent the drawing together, in the action of the Eucharist, of the fruits of nature and the fruits of human work and skill in the total offering of creation. For that reason, Welker suggests, I am sure rightly, that the bread and the wine could never properly be replaced by purely natural products, such as water and apples.[10] The Eucharistic gifts unite nature and human culture. One is reminded of that wonderful picture in Revelation of the worship of heaven, in which the four and twenty elders (representing humanity) and the four living creatures (representing the non-human creation) fall down together in worship and thanksgiving to 'The Lord God the Almighty, who was and is and is to come' (Revelation 4:6-11). I think also of the words of Augustine, Bishop of Hippo, in one of his sermons to his congregation, reminding them that they are there on the altar together with their gifts.

In these ways, the bread and wine that we receive at Com-

10. Ibid., p. 66.

munion are integrated into a profound and all-embracing context, which offers us also some insight into how it is that these gifts become for us truly the means by which we receive the body and blood of Christ. There are certainly mysteries here that are difficult to understand. I want to reaffirm my conviction stated in an earlier writing, that 'In some manner the bread and wine are an integral part of the whole Eucharistic action in a way neither detachably magical nor [simply and] dispensably symbolic'.[11] I am encouraged that the first Anglican–Roman Catholic International Commission, commenting on the statement that the bread and wine become the body and blood of Christ, said,

> *Becoming* does not imply material change. Nor does the liturgical use of the word imply that the bread and wine become Christ's body and blood in such a way that in the eucharistic celebration his presence is limited to the consecrated elements. It does not imply that Christ becomes present in the eucharist in the same manner that he was present in his earthly life. It does not imply that this *becoming* follows the physical laws of this world.[12]

I understand the last point to refer to the way in which we can see the sacrament as involving some anticipation, here and now, of the final eschatological reality of God's new creation, a topic to which I shall return.

The linkage of community and gifts in the single action of the Eucharist is the reason why I personally am unable to share in a certain kind of extra-Eucharistic devotion to the consecrated elements. This kind of 'tabernacle piety', centring on meditation before the reserved sacrament, has been

11. J. C. Polkinghorne, *Science and Providence*, SPCK, 1989, p. 94.
12. ARCIC—*The Final Report*, CTS/SPCK, 1982, p. 21.

subject to some re-evaluation within the Roman Catholic community in recent years. Father Bouyer comments that, in its extreme forms, there had come to be a danger that 'The mass becomes merely a means to refilling the tabernacle'.[13] Of course there is no difficulty about the reservation of the sacrament so that subsequently it can be taken to the sick or the housebound, for this is the way in which they participate with the other worshippers in a single extended Eucharistic celebration.

Sacramental experience is very rich and many-layered, and there is always a danger in any age that the temper of the times will encourage the neglect or distortion of some important aspect of its character. One might suspect that this has happened today in a lack of attention to a dimension of the Holy Communion that was, perhaps, of obsessive concern in Reformation times. I refer to the understanding of the Eucharist as being the place where we receive *the forgiveness of sins and liberation from the power of sin*. We certainly need to recall the gospel insight that, according to Matthew, on the night of his betrayal Jesus spoke of his 'blood of the covenant which is poured out for many for the forgiveness of sins' (Matthew 26:28). Welker detects a degree of tension in Eucharistic theology between this insight and the complementary insight, strongly emphasised in the Johannine teaching, that looks to the dimension of eschatological hope expressed in the words of Jesus, 'Those who eat my flesh and drink my blood have eternal life and I will raise them up at the last day' (John 6:54). Welker regards a concern with sins forgiven as more characteristic of Reformation sacramental theology and a concern

13. Bouyer, *Eucharist*, p. 10.

with being sustained in eternal life as more characteristic of Catholic thought. He links the theme of liberation from the power of sin with the significance that he attributes to the context of compromise and threat in which the sacrament was instituted and in which it continues to be celebrated.[14]

The acknowledgement of human need for deliverance from sin is important and it can act as an antidote to too easy and unthinking an approach to the Holy Mysteries. Yet, many Reformation liturgies expressed an obsessive preoccupation with the issue of sin and its remedy, creating an imbalance that reflected the particular salvific concerns of the time, which centred on how it could be that sinful human beings received acceptance from a holy God. The Anglican Book of Common Prayer of 1662, drawing much of its material from the work of Thomas Cranmer more than a century earlier, is a good example of this tendency. Its Communion liturgy has a heavy emphasis on penitential material and on the proferring of reassurance to those who are conscious of their past failures. Confession and absolution are immediately followed by the further reassurance of the 'Comfortable Words', which in turn are followed by a further Prayer of Humble Access, before the worshipper is considered to be ready to approach the Communion table. Theologically, the focus is almost entirely on redemption through the cross of Christ. One of the major driving forces in producing modern liturgies within Anglicanism has been so that they can offer a better balance between cross and resurrection in their articulation of the gospel, and so express both the forgiveness of sins and the gift of eternal life, in a way that did not prove possible for Cranmer in his

14. Welker, *What Happens at Holy Communion*, ch. 10.

response to the spiritual atmosphere of the sixteenth century. One of the gifts that we have received from the Reformers, for which we should be grateful, is the recognition that the benefits that flow from Christ are not given us quasi-magically or mechanically, *ex opere operato*, but have to be received fittingly by the faithful as they participate in the total action of the Lord's Supper.

Further important dimensions of the Eucharist demand our attention. The first of these is that, though the Eucharist is celebrated on each occasion at a particular time and in a particular place, nevertheless it has a universal character that brings into focus all places and *the plenitude of times, past, present and still to come.* In many modern liturgies, the whole congregation, in the course of the great prayer of thanksgiving, is called on to affirm that 'Christ has died, Christ is risen, Christ will come again'. The Eucharist is celebrated now, in the presence of the risen Christ who is ever contemporary, while at the same time it looks back to the events of Jesus' earthly life, in particular to his sacrificial death on Calvary, and it looks forward expectantly to the hope of the eschatological future, already beginning to come into being through the seminal event of Christ's resurrection from the dead, and eventually to be fulfilled by the visible vindication of his Lordship, of which the Second Coming is the symbol.

Paul told the Corinthians that 'as often as you eat the bread and drink the cup, you proclaim the Lord's death until he come' (1 Corinthians 11:26). The aspect of the Eucharist that relates to time past focuses particularly on the cross. The broken bread recalls the Lord's broken body and the poured-out wine recalls his shed blood. These 'remembrances of him' are something very much more powerful than simple historical reminiscence. The Greek word translated 'memory' is *anamnesis*, and it can carry the force of an event of the past that is re-presented now, for contemporary participation in it, as when Jewish people at Passover still share in the Exodus event of all those centuries ago. Similarly, the Christian worshipper at Holy Communion participates in the sacrificial event of Christ's death on the cross. In the twenty-first century, sacrificial language is difficult for many people, but it has always played an important role in sacramental theology. Today there is virtually universal recognition that the Mass is not a re-enactment or simple continuation of Christ's once-for-all death on our behalf at Calvary, but in the Eucharist there is an anamnetic re-presentation of that sacrifice, enabling us to participate truly in that sacrificial event, as we offer ourselves to the Father through the Son and receive the benefits of Christ's passion through the Holy Spirit's work within us, in the doxological context of worshipful praise.

The aspect of the Eucharist relating to the present is, of course, that the risen Lord is indeed with the worshippers. In my next chapter, when I turn to eschatology, I shall try to say something about the worlds of the old creation and of the new. I believe that Christ's resurrection, occurring within the history of this present world but going beyond the confines of that world, implies that these two worlds are not simply sequential but are related to each other in very subtle ways. I want to suggest that the time of this world and the 'time' of the world to come exist, so to speak, alongside each other, neither totally separated nor simply coincident. One way of understanding Eucharistic experience is to believe that in the covenanted occasions of the sacrament these two worlds draw closer together than they are at ordinary times. We can see here the work of the Spirit, participating in the sacramental action and acting as the 'first instalment' (*arrabon*: 2 Corinthians 1:24) of the life to come in Christ.

The future-oriented aspect of the Eucharist expresses what one might call the interim character of a sacrament that proclaims 'the Lord's death *until he come*' (1 Corinthians 11:26). The future aspect also relates to the profound insight, expressed within history and best preserved in the Orthodox tradition, that Eucharistic worship here and now is already a participation in the unending worship of the heavenly host that fully awaits us beyond our deaths. The congregation gathered on Earth is joined spiritually to the four and twenty elders and the four living creatures who fall down before the throne of God and of the Lamb. It is indeed 'with angels and archangels and with all the company of heaven' that we sing the Sanctus.

The expectation of Christ's *parousia*, his coming again, is a sharing in the hope that God's kingdom will indeed come, so that Holy Communion is a preparation for that consummating event in which the Lordship of Christ will be fully and visibly set forth, and God will in truth be 'all in all' (1 Corinthians 15:28). Sacramental experience in this world is a foretaste of the life awaiting us in a world that will be totally sacramental in its character, wholly suffused with the life and energies of God.

A powerful eschatological image is that of the Messianic Banquet (Isaiah 25:6). It is a pity that we have almost lost the understanding that the Lord's Supper has the form of a communal meal, for this would enable us to see the sacrament as anticipating that symbol of the final great gathering of the Lord's redeemed. Liturgical practices that encourage the recognition of shared participation in the sacrament, as when the communicants gather round the table to receive as a body, rather than as a succession of individuals, help to reaffirm this insight.

The final aspect of the Eucharist that we must consider is its *Trinitarian character*. All Christian prayer is made to the Father, through the Son, and in the Spirit. In Holy Communion, this Trinitarian basis is of particular importance. Modern liturgies, partly through the recovery of ancient forms such as that recorded by Hippolytus in the *Apostolic Tradition*, have made this Trinitarian centrality much clearer than was the case in many of the liturgies stemming from the Reformation. The great prayer of thanksgiving now usually begins with praise to the Father for the work of creation. This is the point at which the sacrament's Eucharistic character is most clearly displayed in the thanksgiving (*eucharistia*) of the Christian community. The liturgy will then move to a representation (*anamnesis*) of the work of the Son, in the redeeming sacrifice of the cross and the resurrection from the grave, which brings about the forgiveness of sins and deliverance from the power of evil and mortality. At some point, most modern liturgies also include an *epiclesis*, or calling down of the Holy Spirit, frequently expressed in a twofold way, invoking the Spirit's descent both on the gathered worshippers and also upon their gifts. The essence of sacramental action is its joining in a complementary relationship both the material and the spiritual aspects of creation, so that the community of the faithful is sanctified together with its offerings of bread and wine. In an important way this complementarity emphasises that the Spirit relates to the whole of creation. It is interesting to note that the introduction of the *epiclesis* into the ancient liturgies seems to have taken place in the second half of the fourth century, just the time when, after much theological struggle, the Church attained a clear recognition of the status of the Holy Spirit as the Third Person of the Holy Trinity. In the liturgy of Holy Communion we see how the *lex orandi* (the order of prayer) and the *lex credendi* (the order of belief) are harmoniously integrated, as the worshippers present at the sacrament are drawn into the grand Christian themes of creation, redemption and sanctification.

Our exploration of Eucharistic theology has illustrated to some extent the depth and subtlety of the kind of interpreted experience that constitutes the motivation for Christian beliefs. Sacramental theology is as complex and sophisticated, and ultimately as powerfully insightful, as the considerations that support a fundamental theory in science. In neither case would one expect the lines of argument to be superficially simple or naïvely accessible. In each case, the cost of illumination is the willingness to have one's everyday habits of thought revised and expanded under the influence of the reality encountered.

This chapter is a little shorter than the others in this book. I have to confess that I have exhausted all that I am able to say about these Holy Mysteries which lie at the heart of my own Christian life. It is characteristic of the way in which scientists think that when we feel we have said all that we can, we just stop and do not try to elaborate further. I simply want to emphasise the fact, again so familiar to a scientist, that an inability to form a totally articulated interpretation of all aspects of a realm of experience is never grounds for the foolish denial of the reality of that experience. Perhaps I can leave the last word to someone who, if not exactly an Anglican divine, did have a great deal of influence on the formation of Anglicanism. Queen Elizabeth I wrote:

'Twas God the word that spake it,
He took the bread and brake it;
And what that word doth make it,
That I believe and take it.[15]

Amen to that.

---

15. *Oxford Dictionary of Quotations*, 2nd edn, Oxford University Press, 1953, p. 197.

CHAPTER SIX

Eschatological Exploration

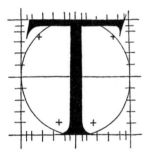

HE great Elizabethan seaman Sir Francis Drake once said that 'There must be a beginning of any great matter, but the continuing until the end until it be thoroughly finished yields the true glory'. If we apply this test of ultimate completion to the Creator's work in the existing universe, the verdict is one that presents a challenge to theological understanding. Life on Earth will certainly not continue for ever. Even if it survives the hazards of future local catastrophes, either of a natural kind (such as the asteroid impact that eliminated the dinosaurs sixty-five million years ago) or of the kind that results from human folly (such as an all-out nuclear war), there will inevitably come a time when the Sun will have used up all its core hydrogen fuel, with the consequence that it will expand to form a red giant, thereby destroying any life still surviving on this planet. This solar disaster will not happen for a very long time, for the Sun has about five billion years' worth of hydrogen still

143

left to burn, but it is certain to happen one day. Of course, by then life may well have migrated elsewhere in the galaxy but, from the point of view of the longest significant time-scales, that would amount to no more than a temporary reprieve. Ultimately the whole universe is condemned to a final futility, either as a result of the bang of collapse back into the Big Crunch or as the result of the whimper of decay into low-grade radiation, expanding and cooling for ever. Such are the reliable prognostications that the cosmologists can make on the basis of extrapolating the present state of physical process into its final stages. If things continue as they have been, it is as sure as can be that all forms of carbon-based life will prove to have been no more than a transient episode in the history of the universe. How then can theology claim to interpret this dismal forecast as the 'thorough finishing' of God's creation?

Here is the eschatological challenge that science presents to theology. It raises the question of whether the creation really is a cosmos at all, or ultimately just a meaningless chaos. Does cosmic history make complete sense, or will the story eventually prove to have been that of the senseless succession of one thing after another? Most contemporary theologians appear to be untroubled by the issue and to give no serious attention to the questions that it raises. The temporal horizon of their thinking is confined to the short span of known human cultural history and to its conjectured continuation a few thousands of years into the future. Such temporal myopia is unacceptable in a subject that seeks to deal with fundamental concerns. Fortunately it has not been the stance adopted by all theologians. A few years ago a group of scholars, drawn from both theology and the sciences, was assembled at the Center of Theological Inquiry at Princeton precisely in order to work together on eschatological matters viewed in the widest possible perspective. The results of the collaboration represented by this CTI Eschatology Group were eventually published in a joint volume, *The End of the World and the Ends of God*.[1] In that book we declared our shared conviction that 'it is of the highest importance that Christians and the Christian Church should not lose nerve in witnessing to our generation about the eschatological hope that is set before us',[2] and we sought to conduct our thinking in the light of all that science, both natural and human, could tell us that was of relevance to our task. One can see the work of the Group as marking a further drawing together of science and theology in their developing mutual conversation. The project formed the next step in a long-continuing process by which that interdisciplinary interaction has spiralled inwards, moving on from creation and natural theology, through an intense period of the discussion of divine action, to eschatological exploration, where theological considerations play a yet more dominant role. In this last phase, the role of science has been principally to pose some of the questions to which theology has then had to seek answers from its own resources. Yet we shall see that it is also the case that scientific insights have a second role, in motivating the imposition of certain metaphysical constraints on what could be taken to be a satisfactory theological response.

The experience of working in the CTI Eschatology Group led me to further writing on eschatological issues.[3] At

---

1. J. C. Polkinghorne and M. Welker (eds.), *The End of the World and the Ends of God*, Trinity Press International, 2000.

2. Ibid., p. 13.

3. J. C. Polkinghorne, *The God of Hope and the End of the World*, SPCK/Yale University Press, 2002.

the close of my book that resulted from this work, I found that I could identify four 'eschatological criteria' whose fulfilment seemed to be essential for a credible theology. I wish to organise the discussion of this chapter under the corresponding four heads.

(1) *If the universe is a creation, it must make sense everlastingly and so ultimately it must be redeemed from transience and decay.*[4]

This is the nub of theology's response to the scientific prediction of cosmic collapse or decay. It is not for theology to deny the validity of science's 'horizontal' extrapolation of present physical process, but theology can point out the limited character of the prediction, in that it necessarily fails to take into account the 'vertical' resources, not accessible to scientific thinking, that point us beyond an ultimate cosmic dying to the eternal reality of a faithful Creator. What is inconceivable within a narrow perspective may become conceivable when the metaphysical horizon is widened. What the scientific scenario does demonstrate, however, is the ultimate bankruptcy of a naturalistic evolutionary optimism. The unfolding of present process alone will not bring about a golden age, for its history must peter out in final futility.

As I stated earlier, I do not believe that the problem posed by our knowledge of the universe's eventual death on a timescale of very many tens of billions of years gives rise to theological perplexities any more intense than those raised by the even more certain knowledge of our own deaths on a timescale of tens of years. Jesus dealt with the latter in his discussion with the Sadducees (Mark 12:18–27). In response to their ques-

4. Ibid., p. 148.

tioning of the idea of a destiny beyond death, he took them back to Exodus where, at the burning bush, God is declared to be 'the God of Abraham, the God of Isaac and the God of Jacob' (Exodus 3, *passim*). Jesus comments, 'He is the God not of the dead, but of the living; you are quite wrong' (v. 27). The logic is incontestable if one accepts that there is a God who is everlastingly steadfast and faithful. If the patriarchs mattered to God once—and they certainly did—then they must matter to God for ever. If we matter to God once—and we certainly do—then we must matter to God for ever. The faithful Creator will neither abandon creatures nor rest content simply with a perfect recall in the divine memory of what they have already been. There is a sense of incompleteness about the span of life in this world that renders inadequate the idea of simply the preservation of what has been, and which demands a continuation of human lives beyond death into the further possibilities of what they might become. Persons must really continue to live before God and so attain the 'true glory' of the unending fulfilment of eternal life.

Many theologians are able to assent to this eschatological insight in relation to the destinies of human beings, but the considerable anthropocentricity of so much theological thinking has made it difficult for many of them to appreciate fully the necessarily cosmic scope of the Creator's total concern. This vast universe is not just there to be the backdrop for the human drama, now taking place after an overture that has lasted fourteen billion years. It all has a value of its own. All creatures must be of concern to their Creator, in appropriate ways. Some theologians have been able to grasp this in relation to the planetary home of humanity. An ecotheology will recognise that the Earth and its biological inhabitants must in some way attain their fulfilment and liberation from the shackles of threatened futility. Yet this hope is too often conceived in terms of some kind of green revolution which, however desirable it might be in itself, would only be as transient as any other episode in the history of carbon-based life.

Comparatively few contemporary theologians seem capable of taking the universe itself with a sufficient degree of creaturely seriousness. It is remarkable, therefore, that the writers of the New Testament were able to overcome such parochiality, albeit while expressing themselves in terms of the cosmological understandings of their day. We read in Colossians of the cosmic significance of Christ, the one through whom God 'was pleased to reconcile to himself all things [*ta panta*, not just *hoi pantes*, all people], by making peace through the blood of his cross' (Colossians 1:20). Most striking of all is the well-known passage in the middle of chapter 8 of Romans, in which Paul writes of a creation that 'was subjected to futility, not of its own will but of the will of the one who subjected it, in hope that the creation itself will be set free from its bondage to decay, and will obtain the freedom of the glory of the children of God' (Romans 8:20-21).

Our contemporary understanding of the nature of evolutionary process makes it easier for us to understand the 'bondage to decay' that characterises the present universe. Transience is essential in an evolving world, which can only explore and realise its inherent fruitfulness through a process in which the generations inexorably give way to each other. In such a world, death is the absolutely necessary cost of new life. This kind of world is what Paul elsewhere calls the 'old creation', in comparison to the 'new creation' that ultimately will be its redeemed transform (2 Corinthians 5:17; see also Revelation

21:5). We have seen that we can understand an evolving world theologically as being a creation given by its Creator the freedom to 'make itself', thus bringing its potentiality to birth in its own way (p. 67). That kind of world is a great good but, because of its restlessly changing nature, it can never be the home of an everlasting good. Thus we are led to think of creation as being an intrinsically two-step process: first the old creation making itself in the context of change and decay, and then the new creation that will no longer be in bondage to decay. These two stages of creation are linked together by the new arising not *ex nihilo* but *ex vetere,* as the resurrected and redeemed transformation of the old. The character of this new creation is something to which we shall have to give further attention as the eschatological argument develops.

out of nothing
out of old

(2) *If human beings are creatures loved by their Creator, they must have a destiny beyond their deaths. Every generation must participate equally in that destiny, in which it will receive the healing of its hurts and the restoration of its integrity, thereby participating for itself in the ultimate fulfilment of the divine purpose.*[5]

Accepting the validity of this criterion amounts to a denial of any concept of ultimate utopia fully attainable within present history. It shows that true hope cannot centre simply on the achievement of some this-worldly state of affairs, though that recognition should by no means discourage the continuing human struggle for peace and justice on Earth. Were such a terrestrial kingdom to come, it could only be the subject of the transient enjoyment of its mortal members, and it would have been denied totally to all the generations that had preceded its coming. I do not believe that this could

5. Ibid.

amount to sufficient fulfilment to make total sense of creation or to be a fitting realisation of God's final purposes.

If we look for a scriptural basis for belief that this second criterion will be satisfied, we can find it in Jesus' words about Abraham, Isaac and Jacob and his proclamation that God is not the God of the dead but of the living. Here is the basis for a hope for everyone. We shall all die with our lives incomplete, with possibilities unfulfilled and with hurts unhealed. This will be true even of those fortunate enough to die peacefully in honoured old age. How much more must it be true of those who die prematurely and painfully, through disease, famine, war or neglect. If God is the Father of our Lord Jesus Christ, all the generations of oppressed and exploited people must have the prospect of a life beyond death, in which they will receive what was unjustly denied them in this life. Those who died in infancy, and those who died in the death camps, must have life restored to them. That great multitude of people who, at many times and in many places, never had the good news of Christ conveyed to them in a way that they could understand and receive, must have the gospel presented to them in a life beyond the grave. I am not for a moment suggesting a kind of pie-in-the-sky eschatology, offering heavenly life as a compensation for earthly suffering and privation. I do not want to invoke eschatology to 'solve' the problems of theodicy. Those problems remain, but I do say that they would be even less tractable if there were to be no destiny beyond death. The death of the peasant boy in Ivan Karamazov's terrible story, deliberately torn in pieces before his mother's eyes by the hounds set upon him by the general whose dog he had accidently injured, is not 'explained away' by his having a life beyond his awful death, but that death would have been even more terrible if it had led only to the grave. And, I think we must add, there is a deep human instinct that the murderous general must not escape answering for what he has done, even if justice is unable to touch him in this world.

The persistent question that we face in this chapter is, Does the universe make total sense, now and for ever? I have been arguing that the worth of individual human beings—a worth that Christians believe is ultimately bestowed on us by our Creator and Redeemer—is only fittingly affirmed if God is indeed 'the God not of the dead but of the living'. If it is a moral truth that human beings are always ends and never mere means, then it is a theological truth that each individual human being has an everlasting ultimacy in the purposes of the Creator. The Christian way of understanding how that ultimacy is worked out is in terms of death and resurrection, and not in terms of some kind of spiritual survival. What that could actually mean is something that we still have to explore. For now I want to emphasise that Christianity has never denied the reality of death. Gethsemane makes that plain enough, as Jesus struggles with the need to accept the cup of crucifixion that awaits him. It has often been remarked that Jesus' agonised acceptance of his imminent death is in striking contrast to the philosophical tranquillity of Socrates, as described in Plato's *Phaedo*, when he affirms his belief in the immortality of the soul before drinking the cup of hemlock. Death, in Christian understanding, is a real end, but it is not the ultimate end, for only God is ultimate. The last word does not lie with death but with God.

My discussion so far in this section has had the anthropocentric focus that I have criticised in other theologians. It is time to remedy this and to ask what sort of destiny we can imagine for other creatures. Perhaps it is most helpful to fix attention on the question of animal destiny. I believe that every human being that has ever lived will live again beyond their deaths, but should I also believe this to be true of every dinosaur? Even more problematically, what about every bacterium that has ever lived? For two billion years or so, single-celled entities were the only living creatures on Earth and even today they represent a significant fraction of its biomass. Some sort of balanced conjecture seems to be called for. On the one hand, I cannot imagine that there will be no animals in the new creation. That would be an impoverished world. On the other hand, I think it highly unlikely that they will all be there. There is a human intuition, shared by many but not by all, that animals are indeed to be valued, but more in the type than in the token. This is the kind of understanding that enables many of us to agree that it is morally permissible, in circumstances of limited forage, to cull a herd of deer, preserving the group at the cost of the humane killing of some of its members. Such a policy could certainly not be ethically countenanced in relation to a human population. I think it likely, therefore, that there will be horses in the world to come, but not every horse that has ever lived. An intriguing special case is presented by animals who are greatly loved pets. Have they acquired sufficient idiosyncratic significance to require this to be continued beyond death? I do not know. There comes a time when it is best to call a halt to eschatological speculation and to heed the advice, 'Wait and see'.

(3) *In so far as present human imagination can articulate eschatological expectation, it has to do so within the tension between continuity and discontinuity. There must be sufficient continuity to ensure that individuals truly share in the life to come as their resur-*

rected selves and not as new beings given the old names. There must be sufficient discontinuity to ensure that the life to come is free from the suffering and mortality of the old creation.[6]

We come now to the issue that lies at the very heart of any attempt to articulate a credible understanding of the nature of eschatological hope. There is a theological necessity to value the present creation as a world proclaimed by its Creator to be 'very good' (Genesis 1:31). Yet, at the same time one must also recognise that this world in itself is necessarily characterised by transience and, ultimately, by futility. Taken together, these two insights imply that the eschatological transformation that will overcome futility and bring about God's final purposes can be neither just a repetition of this world nor simply an apocalyptic wiping clean of the cosmic slate in order to start all over again with a new world that is radically discontinuous with the old one. The latter would correspond to a divine declaration that the present creation is ultimately valueless and pointless. The image that we have to hold onto as the antidote to apocalyptic pessimism is the fundamental Christian picture of death and resurrection; a real death followed by real and unending new life, in which what had died is restored and transformed in order that it may finally enter into its 'true glory'. Ever since the time when Paul wrote the fifteenth chapter of his first letter to the Corinthians, theology has been struggling to find ways of conceiving the almost inconceivable, as it wrestles with the twin eschatological themes of continuity and discontinuity.

Continuity of a kind is indispensable if it really is to be Abraham, Isaac and Jacob who live again in the Kingdom of

6. Ibid., p. 149.

God. Yet they will have been resurrected and not merely resuscitated, so that discontinuity of a kind is also indispensable. If the new creation were not significantly different from the old creation, people would only live again in order to die again. In our work together, all the members of the CTI Eschatology Group had to struggle with the dialectic of continuity/discontinuity, whatever was the subject of any particular person's contribution to the whole project. It is in relation to the constraints associated with attaining a sufficient degree of continuity that science has something to contribute to eschatological exploration. I would like to consider four conditions of *consistent continuity* that seem to be of importance.

(*a*) *Embodiment.*[7] We have good reason to consider human beings as psychosomatic unities and, therefore, to believe that it is intrinsic to humanity to be embodied. We are not apprentice angels, awaiting release from the entrapment of the flesh, as the gnostic thinkers of the first Christian centuries believed to be the case. This kind of error has proved endemic in Christian thinking, taking different forms of expression at different times. Today, it seems to me, a certain kind of liberal theologian is among those who are most at risk of giving way to undue spiritualising tendencies. Of course, such people do not share the gnostic belief that the material world is the work of an inferior Demiurge who is opposed to the spiritual purposes of the Divine Being, but they do seem content to treat the physical universe as if it were a domain ruled solely by science, with theology lodged in the realm of the spiritual, where it can remain happily invulnerable to whatever detailed consequences arise from scientific discoveries about the physi-

7. Ibid., pp. 104 and 108.

cal. Thus it has become quite popular in some circles to say that God does not interact with nature, but only with human minds. In fact, of course, if human beings are psychosomatic unities, even God cannot interact with a human mind without also interacting with a human brain. What makes things even odder is that many of those who take this view seem also to be very uneasy with a Platonic duality of soul and body. I think that they are right in this respect, but the resulting position seems to me tenable only to the extent that it lacks fully explicit articulation, being allowed to remain fuzzy and unfocused in its metaphysical expression. I fail to be persuaded that this is a helpful way to think.

It is well known that Archbishop William Temple once said that Christianity is the most materialistic of the great world religions. I think that is right, and I am glad that it is so because it expresses an uncompromising realism about what it is to be human. Of course, no more than the archbishop do I mean this in a crassly reductionist sense, as if human beings were no more than complex assemblies of elementary particles. I shall try to explain a little later how I think it is possible to reject both Platonic dualism and physicalist reductionism, and to see human beings, to use a famous phrase, as being animated bodies rather than incarnated souls. Suffice it for the moment to say that in that phrase 'body' is being used in an inclusive (Hebrew) sense to mean the total embodied person. If this understanding of the nature of humanity is correct, then it follows that our destiny beyond death will also be to live in an embodied state. To suppose the contrary would be to settle for a less than human form of future hope. Of course, the life of the world to come will also have to be a transformed kind of bodily existence, if that life is to be everlasting and free from the shackles of mortality. This is the point at which the discontinuity half of the eschatological dialectic must be called into play, a subject to which I shall return. Wrestling with these problems is an ancient Christian preoccupation. As I have already pointed out, in the fifteenth chapter of his first letter to the Corinthians we see Paul engaging precisely with the theme of continuity/discontinuity (vv. 35–50). He uses the mysterious and virtually untranslatable terms *soma psychikon* (which the NRSV renders 'a physical body' - 'natural body' might be a somewhat better choice of phrase) and *soma pneumatikon* ('a spiritual body') to represent what he is talking about, namely the animating principles involved. The repetition of *soma* emphasises the continuity of embodiment, while the differing adjectives emphasise the discontinuity that results from the eschatological transformation involved in resurrection.

(*b*) *Temporality.*[8] Just as it is intrinsic to humanity to be embodied, so it is surely intrinsic to our being that we are temporal creatures. General relativity has taught us that in this universe space, time and matter all belong together in a single indivisible theoretical embrace. You cannot have one without the other two, or even think adequately about them in separation, in the way that Newton had supposed. Matter curves spacetime and the geometry of spacetime curves the paths of matter, so together they constitute a package deal.

It seems likely to me that this kind of integrated relationship will continue to be the Creator's will in the world of the new creation, expressing its coherent unity. If that is the case, then the human destiny beyond death will no more be atem-

8. Ibid., pp. 117–21.

poral than it will be disembodied, though, once again, there will also be a dimension of discontinuity, so that the 'time' of the world to come is not just a prolongation of the time of this world, or simply its immediate successor. Rather, it is a new time altogether, possessing its own independent nature and integrity.

A good deal of theological thinking, both traditional and contemporary, has been as distrustful of time as it has been of materiality. Wolfhart Pannenberg tells us that fulfilment is impossible 'without an end to time'.[9] There has been a strong Christian tendency to picture the consummation of human destiny as consisting in the timeless experience of the beatific vision. I think these views are mistaken. Our destiny is not to share in the atemporality of the divine eternal pole, but to live an everlasting life in the 'time' of the world to come. Behind the ingrained theological suspicion of temporality there hovers the Platonic ghost of the idea that the unchanging is always to be preferred to the changing, that perfection is a static state and not a dynamic process, that being is better than becoming. Of course, in this world time is associated with transience—as the hymn says, 'Time, like an ever-rolling stream, / Bears all its sons away'. But there is no necessary connection between change and decay, between temporality and transience. Here, once again, the requirement of continuity comes into tension with the requirement of discontinuity, and the issue is again one to which we shall have to return later. Meanwhile we need to consider a further condition of continuity that science encourages us to take with very great seriousness.

9. W. Pannenberg, *Systematic Theology 3*, Eerdmans, 1998, p. 587.

(*c*) *Process.*[10] If we learn anything about the character of the Creator from what science can tell us about the history of this creation, it is surely that God is patient and subtle, content to work through unfolding process and not by sudden interventions of arbitrary power. Even the miracle of the resurrection is not a one-off divine tour de force, but the seminal event from which God's ultimate salvific process has started to unfold with implications for all humanity (1 Corinthians 15:22). Theologically it seems entirely appropriate that Love should choose to work through process in this discreet and self-effacing way.[11] We may expect that these divine characteristics will continue to be expressed in the life of the world to come. A number of eschatological insights follow from this recognition.

The first is that God's love and mercy will continue to operate in the new creation as they do in the old creation. They will not be removed from a person at the moment of his or her death. It will not be the case that at the point of death an iron curtain descends, cutting off anyone found on the wrong side of it from any hope of everlasting salvation. God will not say, 'You had your chance over all those years. Now you've lost it and that's that.' Of course, I am not at all saying that the decisions we make, and the beliefs to which we commit ourselves in this life, are not serious matters with important consequences for who we are and who we may become. The more we turn from God in this life, the more difficult and painful it will be to respond to God in the life of the world to come. Yet the divine love will surely not be withdrawn in that world, but will

10. Polkinghorne, *The God of Hope and the End of the World*, pp. 132–4.
11. See J. C. Polkinghorne (ed.), *The Work of Love*, SPCK/Eerdmans, 2001.

continue to seek to draw all people into its orbit. If there will turn out to be those who will resist that love for ever, with its offer of forgiveness and redemption, then they will have condemned themselves to live the life of hell.[12] They will not be in a place of torment, painted red, but in a place of infinite boredom, painted grey, from which the divine life has deliberately been excluded by the choice of its inhabitants. The best imaginative picture that I know of hell is not the torture chambers of Dante's *Inferno*, but the dreary town, lost down a crack in the floor of heaven, of C. S. Lewis's *The Great Divorce*.[13]

The second eschatological insight is that the process of the transforming work of grace will continue in the life of the world to come. We shall all die not only with our lives incomplete and our hurts unhealed, but also with our sins not fully repented of and our hearts not fully made clean. In the brighter light of the new creation we shall begin to see ourselves as we really are and as we are seen by God, and we shall have to come to terms with that painful reality. This is how I understand the serious matter of the judgement to come.[14] I do not envisage it as our appearing before an angry Judge, a kind of celestial Judge Jeffreys, but I see it in the Johannine terms that we shall come to see how often we have preferred darkness to light (John 3:19). The consequences of judgement will not be endless punishment but the hopeful opportunity for purgation. Protestant thinking has been too greatly influenced by the understandable Reformation recoil from medieval distortions of the idea of Purgatory. Yet, if we believe in a God who acts through process and not by overwhelming magic, purga-

---

12. Polkinghorne, *The God of Hope and the End of the World*, pp. 136–68.
13. C. S. Lewis, *The Great Divorce*, Geoffrey Bles, 1946.
14. Polkinghorne, *The God of Hope and the End of the World*, pp. 128–33.

tion will be a necessary element in the unfolding process of salvation. This is a hopeful thought. I am greatly moved by the picture in Dante's *Purgatorio* of sinners toiling up the great Mount of Purgatory that symbolically joins Earth to heaven. As they move from one level to the next, they progressively shed their attachments to the seven deadly sins, and the whole mountain resounds with alleluias each time a person takes that further step towards their heavenly destiny.

The third insight resulting from an acknowledgement of the role of process in the divine plan of salvation relates to the life of heaven itself, for that too will have its character of dynamical perfection.[15] Finite human beings cannot take in the unbounded reality of God in a timeless moment of illumination. Rather, our destiny is the unfolding and unending exploration of the inexhaustible riches of the divine nature that will be progressively unveiled to us. People sometimes say that they would certainly like more life than this world affords, but not a life that goes on without an end. They fear eventual boredom. Even the man who said that when he got to heaven he would play golf every day might eventually come to tire of that pastime. If our future life depended upon our own finite resources, these fears would be justified. But the quality of the life of the world to come depends upon the limitless resources of God, and so it will prove to be a life of unending fulfilment.

(*d*) *Personhood and the soul*.[16] A most important question still remains to be discussed. In fact, it is the central issue in relation to the eschatological condition of continuity. What is it that could be the carrier of the continuity of personhood,

15. ibid., pp. 132–36.
16. Ibid., pp. 103–11.

the link that ensures that it really will be Abraham, Isaac and Jacob who live again in the Kingdom of God, and not just new persons given the old names?

The traditional Christian answer has been that it is the individual human soul. That reply has often been understood in a Platonic sense, with the soul being a spiritual entity, temporarily housed in the present fleshly body but released from it at death. If, as I have asserted, we are to think of embodied human beings as psychosomatic unities, that particular understanding is no longer available to us. Have we then lost the concept of the soul altogether? I do not think so.

Whatever it may actually be, the soul is presumably what one might call 'the real me', the essence of my particular personhood. What this could be is at first sight almost as problematic within this life as it might be beyond death. What is it that makes me today, an elderly balding academic, the same person as the young boy with a shock of black hair in the school photograph of sixty years ago? It certainly is not just a crude material continuity, for that does not really exist. The atoms in our bodies are changing all the time, through wear and tear, eating and drinking. They have been changed many times in the course of our lives. These atoms are not the carriers of human continuity, but the real me is surely constituted by the almost infinitely complex *information-bearing pattern* in which these atoms are organised. This insight accords with a growing understanding in science, initiated by the infant discipline of the study of complex systems, that information (the specification of the dynamical patterns of behaviour) will in future take its place alongside energy as a fundamental category in our understanding of the world in which we live (pp. 82–84). My soul is the pattern that is me.

This is really the recovery in modern dress of an old idea. Both Aristotle and Thomas Aquinas believed that the soul is the form (or pattern) of the body. However, the modern concept of the soul that I am proposing seems to me to differ in some significant respects from its ancient antecedents. It is clear that modern thinking must place a greater emphasis on the soul as having a radically dynamical character, changing as we acquire further experience and more memories. It is something that makes us what we are in our temporality, and not just something that we were given in a timeless way. Our souls develop and grow. This is not to say that there is not an unchanging signature of individual identity that is a part of the soul (our genome would be a component of this), but that there is more to the soul than static being. I also think that if this idea is to prove capable of further development, it will have to incorporate a recognition that the pattern that is me is not simply contained within my skin—it must take into account the rich environment in which I live. It must extend to the web of relationships that play so significant a part in the character of my personhood. Acknowledgement of this highly complex and multi-dimensional character of the soul avoids the rather too dessicated concept that the cool language of 'information-bearing pattern' might otherwise be in danger of suggesting.

One must also be frank enough to acknowledge that at present we are not capable of formulating these ideas with any adequate degree of clarity. This is part of a common human intellectual dilemma. Much metaphysical speculation has to engage with matters that are beyond our powers of full comprehension. Putting the matter bluntly, there has to be a degree of 'hand-waving', and the important thing is to try to wave our hands in the right direction. I think that the concept of the soul as the information-bearing pattern of the body is the right direction in which to pursue our anthropological thinking. I do not believe that we are condemned to a Wittgensteinian silence about that whereof we cannot adequately speak, for it is in the attempt to articulate something at all that we find the beginning of the possibility of understanding. Like Augustine in relation to his writing on the Trinity, we should modestly conclude that it is better to say something than to remain totally silent.

The soul, as I understand it, possesses no intrinsic immortality. The pattern that is me will dissolve at my death, with the decay of the body that carried it. To say so is simply to recognise that we have no naturalistic expectation of a destiny beyond death. That hope can only be given us by a trust in the faithfulness of God. It seems to me to be a perfectly coherent hope to believe that the pattern that is me will be preserved by God at my death and held in the divine memory until God's great eschatological act of resurrection, when that pattern will be re-embodied in the 'matter' of the new creation. Once again we see that a credible Christian hope centres on death and resurrection, and not on spiritual survival.

It is time to turn our attention to the *discontinuity* half of the eschatological dialectic. Here our intellectual resources are almost entirely theological, since the focus of attention is on a world whose characteristics will necessarily be very different from the physical and biological world that is the subject of our present experience and which is the object of the investigations of science. The world to come must be so constituted that its processes are temporal without generating transience, their outcomes fruitful without pain and suffering being their shadow side. In the words of Revelation, it will be a world where God 'will wipe every tear from their eyes. Death will be no more; mourning and crying and pain will be no more, for the first things have passed away' (21:4). It is beyond our powers to imagine the details of such a world, but it clearly seems to me a coherent possibility. The root cause of decay in this present world is the outworking of the second law of thermodynamics, expressing the tendency of closed systems to succumb to randomness. This happens because the possibilities of disorder very greatly outnumber the possibilities of order. In consequence, closed systems drift towards increasing haphazardness. Dissipative systems (such as all living entities) manage to swim against the tide of disorder for a while because the input of energy from the environment, and the export of entropy into it, provide an external resource for maintaining order. One may envisage a new kind of 'matter' endowed with internal organising principles of such power as permanently to overcome any tendency to disorder. If the possibility of such a world is accepted, however, it does raise an acute theological question. If the new creation is to be such an attractive state, why did God bother with the vale of tears of the old creation?

I believe that the answer lies in the recognition, already acknowledged, that God's creative action necessarily has a two-step character. First, in a kenotic act of allowing the creaturely other to possess its divinely granted independence, God brought into being a world that exists at some metaphysical remove from its Creator. The gift given by Love is that creatures should be allowed to be themselves and to make themselves in the veiled presence of God. It is in order to preserve this important kenotic insight that I am unable to accept the notion of a form of presently realised intimate connection between God and creation of the kind that panentheism seems to imply. We have already seen that there are consequences that flow from there being a metaphysical space between Creator and creatures. These include the fact that an evolutionary world, exploring in its own way its endowment of fruitful potentiality, must necessarily be a world of transience and death. The existence of free creatures is a greater good than a world populated by perfectly behaving automata, but that good has the cost of mortality and suffering. Therefore the first stage of creation must be followed by a second stage in which God's good purposes can be fully consummated. It is not the divine will that creation should for ever exist at a distance from its Creator. God's final purpose is that creatures should enjoy fully the experience of the unveiled divine presence, and so share in the divine energies, but this must happen by God's love drawing them, so that it is freely and of their own accord that they enter into the intimacy of that relationship, with the rescue from sin and the liberation into life that it brings. It is this totally grace-filled state that is the life of the new creation. I have already argued that it is to be expected to have its own form of 'space', 'time' and 'matter'. Its process can and will be different from that of this present world, precisely because the world to come will have entered within the veil that now cloaks the divine reality from us. This world contains sacraments; that world will be wholly sacramental. The new creation will be a world wholly suffused with the divine presence. It is entirely rational to believe that its natural process will be of a radically different kind from that which science is able to describe today. I do not accept panentheism as a present theological reality, but I do affirm *the eschatological hope of a sacramental panentheism* as the character of the new creation.

Christians believe that this world of the new creation has already begun to come into being with the resurrection of our Lord Jesus Christ. He is the unique link between the life of God and the life of creatures, the bridge by which entry into the divine life becomes possible for us, so that it is in Christ that eschatological fulfilment will be attained. This is the sense in which Jesus could be represented as saying 'No one comes to the Father except through me' (John 14:6). These words, in my opinion, emphatically do not mean that only those who now—in this life—know Jesus by name can come to be those who know him in his saving fullness in the life of the world to come. God's merciful love will not be limited by historical or geographical contingency.

(4) *The only ground for [eschatological] hope lies in the steadfast love and faithfulness of God, which is testified to by the resurrection of Jesus Christ.*[17]

Much eschatological thinking is necessarily speculative. In advance of actual experience, exploration of possibility and credibility is the best that we can manage and much must remain tentative. Yet, for the Christian our hope has a firm foundation, for it is anchored in the resurrection of Christ. I shall not attempt to argue here for the credibility of that belief in itself, important as I believe it is to do so, because that is a task I

17. Ibid., 149.

have attempted elsewhere.[18] What I am concerned with now is to draw some simple theological conclusions that follow from this belief, which is the point on which so much Christian understanding pivots.

The first thing to be said is that the resurrection affirms the faithfulness of God. On Good Friday that faithfulness must have seemed to be under question, as the one man who wholly committed himself to live a life of faith and trust in his Heavenly Father was seen to undergo an unjust and painful death and to have died with a cry of dereliction on his lips. Hence the deathly silence of Holy Saturday. The resurrection shows that those two dark days, real and undeniable as they were, do not encompass the whole story of God's purposes in Jesus. There is the third day, the day of resurrection, on which it was made manifest that the last word is indeed with God and the divine gift of new life, and not with the power of death.

The second thing is to lay emphasis on the importance of the empty tomb. This is not an altogether popular thing to assert today, since modern belief often tends to find difficulty with any concept of the bodily resurrection of Jesus. Again we see at work a spiritualising tendency that at first sight may seem to make belief easier but which, in the end, proves to be less than adequate for full credibility. If, as I have argued, embodiment is intrinsic to humanity, and if the true humanity of Jesus continues in his risen and glorified life, then he needs to be embodied in that new life. Of course, resurrection is not mere resuscitation. Paul was absolutely right to say that 'flesh and blood will not inherit the kingdom of God, nor does

18. Ibid., ch. 6; J. C. Polkinghorne, *Science and Christian Belief/The Faith of a Physicist*, SPCK/Princeton University Press, 1994, ch. 6.

the perishable inherit the imperishable' (1 Corinthians 15:50). Jesus' risen body is not the same as the body of his earthly life. But it was derived from that earthly body as its glorified trans-mutation through the death-defeating power of God. We en-counter here, in a particularly intense and focused form, the familiar theme of continuity and discontinuity. Jesus' risen body still bears the scars of the passion, but it is also not easy to recognise him at first sight. Almost all the appearance stories centre on a moment of disclosure in which the astonishing truth of who it is suddenly dawns on those present. The risen Christ appears and disappears at will, phenomena that I per-sonally understand as some form of temporary intersection between the worlds of the old and the new creations.

There is a further important insight conveyed to us through the stories of the empty tomb. The fact that the Lord's risen body is the glorified form of his dead body tes-tifies to the fact that in Christ there is a destiny not only for humankind but also for matter, and so for the whole created order. The empty tomb is a sign not only of Christ's resurrec-tion, but also of the cosmic significance of that great event.

And the final thing to say is that, in Christian understand-ing, what happened to Jesus within history is the foretaste and guarantee of what awaits the rest of humanity beyond the end of history: 'for as all die in Adam, so will all be made alive in Christ' (1 Corinthians 15:22). Once again one sees that the res-urrection was not just a divine tour de force, God showing us in a one-off way what divine power can do, but was the in-auguration of the second and ultimate phase of creation, the seed from which an everlasting fulfilment has already begun to grow as the consummation of divine purpose and the satisfaction of human longing. There will indeed be a 'thorough finishing' that will yield true glory.

I would like to conclude this chapter in the same way that I ended my book on eschatology, by saying that 'Christian belief must not lose its nerve about eschatological hope. A credible theology depends upon it and, in turn, a Trinitarian and incarnational theology can assure us of its credibility.'[19]

19. Polkinghorne, *The God of Hope and the End of the World*, p. 149.

CHAPTER SEVEN

Concluding Unscientific Postscript:
In Defence of Particularity

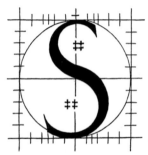

CIENTISTS love generality, and they are often wary of particularity. Professionally we are concerned with explanations based on the laws of nature, which are believed to be the same at all times and in all places. While historical sciences, such as cosmology or evolutionary biology, are indeed concerned with unique sequences of events, the main interest lies in the overall character of what has been going on, rather than in the contingent fine details of actual occurrence. Astrophysicists want to understand the process of galaxy formation, but they do not entertain much ambition to explain the particular way in which the Milky Way acquired the specific structure that we observe. Biologists study the evolutionary history of the horse, but they are content simply to acknowledge a degree of unpredictable contingency in the way one form arose from its predecessor without attributing significance to the precise particularity of the manner in which this occurred.

This preference for the general is one reason why anthropic thinking about the finely tuned specificity of the universe produced so widespread a reaction of shock and scandal in the scientific community. The natural inclination had been to regard the world as no more than just a typical specimen of what a universe might be like. Ideas of a multiverse seem to have become popular just because they represent a way of swallowing up this embarrassing cosmic particularity in the wider generality of a proliferation of different worlds.

A distaste for particularity is an obstacle to serious engagement with religious understanding. The Trinitarian perspective explored in this book must necessarily appeal to the specificities of the Christian tradition, and especially to the scandalous particularity of the incarnation, the belief that God has acted to make the divine nature known most fully in a particular person, living at a particular time and in a particular community. In a recent review, the distinguished physicist Freeman Dyson chided me for the particularity of my Christian perspective.[1] In his opinion, the universe is mysterious, with much that is likely to be significant for its deepest understanding lying beyond the grasp of present human knowledge. For Dyson, the world faith communities maintain ways of living that are oriented towards certain values, and in this sense he is happy to align himself with the Christian tradition. Yet he does not accept—nor appear to pay any attention to—the cognitive claims made by Christianity or by any other faith tradition. He believes that to do so would be to risk surrender to the questionable assertions of the local tribe. Grand

1. F. J. Dyson, 'Science and Religion: No Ends in Sight', *The New York Review of Books*, 28 Mar. 2002.

truth, it seems, is thought to need liberation from any taint of origin in particularity.

Yet when Dyson comes to the exposition of his own thoughts, in fact it is to the particular resource represented by what he calls 'theo-fiction' that he turns for the basis of continuing the discussion. One example of the genre appealed to is Olaf Stapledon's *Star Maker*, a story in which a cosmic quest ends in encounter with the god-like figure of the title, who proves to have no interest in the being or welfare of puny human individuals. One can read such a text as an exploration of the strangeness and otherness of deity, together with an expression of the human fear that ultimately the universe might prove to be hostile or tricksy in relation to ourselves. Such darker feelings are not totally foreign to the religious traditions, for questionings of this kind are to be found in the spiritual writers, for example in some of the Psalms and in the book of Job. I believe that in the biblical tradition such thoughts are treated more profoundly than they are in the tales of theo-fiction, and they are held in tension with positive insights of the fruitfulness of creation and the dignity of humankind, issuing ultimately in an affirmation of trust in the steadfast faithfulness of God. The notion of some degree of cosmic meaningfulness is not altogether unfamiliar to Dyson, as he somewhat guardedly articulated in his often-quoted remark, 'The more I examine the universe and the details of its architecture, the more evidence I find that the universe in some sense must have known that we were coming'.[2]

Rejection of religious particularity may well be re-

2. F. J. Dyson, *Disturbing the Universe*, Harper & Row, 1979, p. 256.

inforced in a scientist's mind by the contrast between the remarkable unanimity of assent that a well-winnowed scientific theory can command and the diversity of claims made by the faith traditions. We are surely right to be impressed by this universal aspect of science, but it would be unreasonable to demand that all forms of human enquiry conform to the scientific pattern. Science purchases its success precisely by bracketing out idiosyncrasy, restricting its consideration to those impersonal modes of encounter with reality that are sufficiently abstracted in the character of what is relevant as to be in essence repeatable, a property that they possess by virtue of their independence of collateral detail, such as the personality of the experimenter.

Many things can be understood in this general way, but by no means all. Of the many events known to us to have occurred in the course of the fourteen-billion-year history of the universe, one of the most remarkable is surely the coming to be of persons on planet Earth. In humanity the universe became aware of itself and, as a by-product of this reflexivity, science itself became an eventual possibility. Pascal said of human beings that, frail reeds though they appear to be on the cosmic scale, nevertheless they are greater than all the stars, for we know them and ourselves and they know nothing.[3] Many of us cannot believe that this emergence of personhood is simply a happy but meaningless accident, for it seems to be a most important clue to the nature of the reality within which we live. There is an authenticity in personal experience, both individual and communal, that demands to be treated with the utmost seriousness. Yet personal experience is always

3. B. Pascal, *Pensées* (1670), frag. 347.

particular experience; it is always a view from somewhere and never a view from nowhere or from everywhere. One of the prime ways to explore personhood is through literature, and the greatest literature is concerned with particular people and particular experiences, rather than the abstracted conceptual figures of a generalised humanity. Its concern is with Hamlet and Lear, and not with Everyman. In his own argument Dyson turned to specific theo-*fiction* for inspiration. In the realm of the personal one cannot reject on a priori grounds the possibility also of the significance of theo-*history*, the actually enacted and particular story of the self-disclosure of profound aspects of ultimate reality. Moreover, in the transpersonal realm of the theological, in contrast to the impersonal realm of science, the initiative for search and discovery does not have to come solely from the human side. It is a coherent possibility that God has acted in specific ways to make the divine nature known and to draw people to that knowledge.

Christianity, of course, is based on just such a claim of a true theo-history. Christian thinking cannot be rid of the scandal of particularity, as it affirms the divinely chosen role of Israel and the unique role of Jesus Christ. These are indispensable parts of its canonical character. The rational way to approach this issue is not with the prior certainty that the universe must remain mysterious because, if it had a deep meaning, this could only be found by means of quasi-scientific generality and not through personal specificity. Instead, we have to enquire whether there is indeed sufficient motivation to embrace such a difficult but exciting kind of particular belief. In this book, and in a somewhat different and more extensive way in my Gifford Lectures,[4] I have sought to contribute something to this enquiry into the credibility of Christianity. My personal conclusion is supportive of the particularity of Christian belief.

But is not the real problem that there are too many competing theo-histories on offer? In contrast to Christianity, Judaism locates the specificity of divine disclosure in the pattern of a particular way of life (*Torah*), while Islam looks to the particularity of a specific divinely dictated book (the Qur'an). When we add to the already diverse assertions of the Abrahamic faiths the different insights of the Eastern religions, the confusion of cognitive claims seems to increase considerably. Of course, the variety of different cultural viewpoints and experiences involved is clearly a source of some of this diversity of expression. Personally, however, I do not find it possible to ascribe all the clash of conflicting convictions to the effects of plural historical and geographical factors. Is the human person of unique and persisting significance (the Abrahamic faiths), or recycled through reincarnation (Hinduism), or ultimately an illusion from which to seek release (Buddhism)? These conflicting concepts do not seem to be culturally different ways of expressing the same idea.

The problems presented by religious diversity are serious, and they are only partly counterbalanced by the acknowledgement that beneath the bewildering diversity one can also discern a degree of commonality of encounter with that dimension of reality called the sacred. The presence of an element of religious unanimity is most clearly demonstrated by

4. J. C. Polkinghorne, *Science and Christian Belief/The Faith of a Physicist*, SPCK/Princeton University Press, 1994.

the similarity of testimony given by the mystics of all traditions when they describe their unitive experiences. (As a Christian, I understand this mystical common denominator as arising from encounter with God in the mode of divine immanence.)

One may hope that dialogue between the faith traditions, only just beginning to take place with due seriousness and probably needing centuries rather than years for its full development, will help to resolve some of these perplexities. We need a way of proceeding that neither denies that all traditions preserve true reports of encounter with sacred reality nor seeks to smooth over their individual differences.[5] I feel that I have to approach such a dialogue from the perspective of my Trinitarian belief.[6] To do otherwise would be disingenuous, and I expect my brothers and sisters in other faith traditions also to speak from where they are.

I certainly do not believe that the answer to these problems lies in the abstracted cosmic agnosticism of a Kantian kind that Dyson seems to advocate. In itself, that approach constitutes a particular point of view, one that simply averts its eyes from the rich, if perplexingly varied, testimonies of the faith traditions (rather as many physicists in the period 1900–25 averted their eyes from the puzzling paradoxes and cognitive clashes that emerging quantum phenomena seemed to present). I do not think that so cautiously detached a strategy is the way to make progress, for it deliberately abandons a good

---

5. An approach of this kind is Keith Ward's programme of comparative theology: see especially his *Religion and Revelation, Religion and Creation, Religion and Human Nature,* and *Religion and Community,* Oxford University Press, 1994, 1996, 1998 and 2000.

6. Polkinghorne, *Science and Christian Belief/The Faith of a Physicist,* ch. 10; *Science and Theology,* SPCK/Fortress Press, 1998, ch. 7.

deal of evidence and insight that needs to be taken into account, however demanding that task might be. Nor do I think that progress lies in the direction of bland lowest-common-denominator formulations, so rarefied in content that virtually no adherent of any faith tradition would consider them worthy of assenting to, or even worth arguing about.

There is another kind of particularity on which the discussion has so far relied, which also needs assessment. It is the acceptance of the specific significance of contemporary terrestrial humanity. What if there are intelligent beings elsewhere in the universe, perhaps possessing powers of understanding orders of magnitude beyond that which humans are capable of attaining? If that were the case, then surely they, rather than us, would be the ones whose thoughts were worth anything in relation to these deep matters. Science does not really know what to think about this extraterrestrial possibility. There must surely be many sites in the universe suitable for the development of some form of life, but until we understand the biochemical pathways by which this has happened on Earth, we do not know how likely it is actually to occur elsewhere. Distinguished scientists disagree radically about this,[7] so we may conclude that the question remains unsettled. Theology does not altogether know what to think about extraterrestrial possibilities either. God's creative purposes may well include 'little green men' as well as humans, and if they need redemption we may well think that the Word would take little green flesh just as we believe the Word took our flesh. If we ever do make contact with intelligent life elsewhere, that will

7. See, for example, F. Crick, *Life Itself*, Macdonald, 1982; C. de Duve, *Vital Dust*, HarperCollins, 1995.

be of great significance both scientifically (do they have the same genetic code as us?) and theologically (what are their religious experiences and beliefs?). In the mean time, it seems difficult to draw any useful conclusions from so speculative a realm of thought.

Perhaps a better-founded notion would be to look not elsewhere in space, but towards the future here on Earth. The average life of a species is just a few million years. May not *homo sapiens* eventually evolve into a 'higher' form of life, qualitatively different from us and with intellectual powers greatly exceeding those that we can attain? The emergence of such beings could represent as great an advance on us as did the appearance of our self-conscious hominid ancestors in relation to their predecessors. In that case, it would surely be these superbeings yet to come who will be equipped adequately to pursue the profound questions of cosmic significance. Are we not in danger of being presumptuously premature in our thinking, unjustifiably inflating the significance of our current form of particularity? Such questionings should not be allowed to induce intellectual paralysis. No one knows what the far future might hold, but this should not deter us from making use of the resources of the present—the only ones, in fact, available to us. Experience gives us good reason to believe that human abilities to attain understanding, while not unlimited, are of a kind to yield worthwhile gains in both science and theology.

As a Christian who believes that God took human life in Jesus Christ, I inevitably think that there must be something of particular significance in present humanity. Moreover, extrapolation into a hypothesised future is far from straightforward. The coming to be of culture, with its Lamarckian power to transmit information directly from one generation to the next and with a compassionate concern to prevent the weak from simply perishing in a struggle for survival, has profoundly modified the evolutionary process in ways that we do not know how to evaluate. In addition, while terrestrial history has seen a number of stages of the emergence of radical novelty—life, consciousness, self-consciousness—there is no clear reason to believe that there has to be a next step waiting in the sequence. It may be that the degree of complexity represented by the human brain is close to a limiting case in the fruitfully organised properties of individual material systems.[8]

Before finally leaving the issue of particularity, I want to say one more thing. I believe that particularity is of significance in relation to philosophy as well as in relation to theology. That might seem a strange assertion to make, since the aim of the philosophical discourse of modernity seemed precisely to be the attainment of general validity. While this is clearly desirable where it can be had—and I do not want to give way to a postmodern mood of epistemological despair— its pursuit should not blind us to the possibility that we need also to take into account a role for contextual particularity. In making this remark, I have in mind the philosophy of science. At the end of all the intensive and varied discussions of the twentieth century, we are left with the dilemma that, on the one hand, there is broad agreement that the scientific method does not rest on unquestionable foundations, for the inextricable intertwining of theory and experiment, experience and interpretation, introduces an element of apparently precari-

8. Perhaps qualitatively new developments would need to be communitarian rather than individualistic, an idea explored by Pierre Teilhard de Chardin in *The Phenomenon of Man*, Collins, 1959.

ous circularity into the argument, while, on the other hand, the advance of scientific understanding in a well-winnowed regime has such a character of the attainment of verisimilitude that merely positivistic or pragmatic accounts seem to fall significantly short of describing what is actually going on. I believe that the resolution of these tensions lies along theological lines. I do not believe that the successful pursuit of scientific discovery is an activity that is feasible in every conceivable world, but it is a fact of our experience in this particular world. By that I mean, as I suggested in Chapter 3, that the explanation of the success of science in exploring the intelligible universe is ultimately theological rather than philosophical, believing as I do that it derives from the fact that this specific universe is a creation endowed with a rational order that is accessible to creatures who are made in the image of the Creator, rather than deriving from general human rational powers that could be exercised equally in any kind of world.

Index